KNOW & GROW
VEGETABLES

KNOW & GROW VEGETABLES

P. J. Salter, J. K. A. Bleasdale,
and others

Oxford New York Toronto Melbourne
Oxford University Press · 1979

Furthermore, most of these excellent masters of the art have only ever tried to paint in certain limited ways. They try some idea to see if it improves things and if in the year that they tried it they think things were better it becomes part of their technique for producing a masterpiece. Scientific research requires to know that any innovation does really make a difference. Careful comparisons are made, often at sites with different weather and soil conditions, and success is claimed only when the conditions required to obtain a consistent result have been established. Furthermore, the scientific approach requires that some understanding of the processes and events that lead to the result should be obtained. This understanding helps in deciding the other conditions under which a given result can be usefully applied.

This book aims to give you not only the results of research but also that understanding of the underlying principles which will help you to apply them. The authors believe that a proper understanding will not only produce for you better results but it will also increase the satisfaction which is so much a part of successful vegetable culture. The emphasis in the book is always to present the practical results of research in a manner that the vegetable gardener will readily understand and will equally readily be able to use.

The authors, who are all members of the staff of the National Vegetable Research Station at Wellesbourne in Warwickshire, U.K. have a comprehensive first-hand knowledge of the research about which they are writing. All of them are enthusiastic gardeners who have had considerable experience at passing their research knowledge on to gardeners. They and their colleagues started doing this in a series of Gardeners' Leaflets which have been widely acclaimed as a breakthrough in gardening advice.

The chapter headings give a clear indication of the subjects covered. It will be apparent that this book is not intended to be a comprehensive guide to vegetable growing, but each subject included is dealt with in depth and at the end of each chapter the conclusions are summarized in tabular form for easy reference.

KNOW & GROW
VEGETABLES

P. J. Salter, J. K. A. Bleasdale,
and others

Oxford New York Toronto Melbourne
Oxford University Press · 1979

Oxford University Press, Walton Street, Oxford OX2 6DP

OXFORD LONDON GLASGOW
NEW YORK TORONTO MELBOURNE WELLINGTON
KUALA LUMPAR SINGAPORE JAKARTA HONG KONG TOKYO
DELHI BOMBAY CALCUTTA MADRAS KARACHI
IBADAN NAIROBI DAR ES SALAAM CAPE TOWN

British Library Cataloguing in Publication Data
Know and grow vegetables.
1. Vegetable gardening
I. Salter, Patrick Jeremy II. Bleasdale, John Kenneth Anthony
635 SB322 78-41123
ISBN 0-19-857547-5
ISBN 0-19-857563-7 Pbk

Printed and bound in Great Britain by
Richard Clay (The Chaucer Press) Ltd, Bungay, Suffolk

Preface

The application of the results of research, mostly carried out within the last twenty-five years, has transformed commercial vegetable production in many parts of the world. Yields of some crops have more than doubled and highly efficient systems of production ensure that the crops are grown to meet stringent market requirements, which include freedom from blemishes caused by pests and diseases. These advances are the direct result of knowing and understanding more about how vegetables grow, how they respond to different weather conditions and cultural operations, and how their pests and diseases can be controlled. Many of the research findings are just as applicable in the garden as they have been shown to be in large-scale production. Indeed some of the research has produced results more applicable in the garden than in the field. This sometimes happens because the research scientist, particularly in the early phases of his work, often works with small plots of vegetables, rather like the gardener, and has to develop techniques suitable for this small scale. This book presents some of the conclusions arrived at as a result of modern research. They are presented in a manner which reveals their practical value to the vegetable gardener.

The benefits to be derived from applying the information in this book are certain and considerable. The essence of science is that its observations should be repeatable. Thus, if it works for scientists it will, with a high degree of certainty, work for you. Gardening is often considered to be an art where the skills of the gardener are of paramount importance. Indeed, most gardening books are written by the master artists of the gardening world who can themselves produce masterpieces. Some of them are very good at describing how they 'paint' their masterpieces, but the majority of vegetable gardeners have neither the time nor the resources or natural skill to emulate their example.

Furthermore, most of these excellent masters of the art have only ever tried to paint in certain limited ways. They try some idea to see if it improves things and if in the year that they tried it they think things were better it becomes part of their technique for producing a masterpiece. Scientific research requires to know that any innovation does really make a difference. Careful comparisons are made, often at sites with different weather and soil conditions, and success is claimed only when the conditions required to obtain a consistent result have been established. Furthermore, the scientific approach requires that some understanding of the processes and events that lead to the result should be obtained. This understanding helps in deciding the other conditions under which a given result can be usefully applied.

This book aims to give you not only the results of research but also that understanding of the underlying principles which will help you to apply them. The authors believe that a proper understanding will not only produce for you better results but it will also increase the satisfaction which is so much a part of successful vegetable culture. The emphasis in the book is always to present the practical results of research in a manner that the vegetable gardener will readily understand and will equally readily be able to use.

The authors, who are all members of the staff of the National Vegetable Research Station at Wellesbourne in Warwickshire, U.K. have a comprehensive first-hand knowledge of the research about which they are writing. All of them are enthusiastic gardeners who have had considerable experience at passing their research knowledge on to gardeners. They and their colleagues started doing this in a series of Gardeners' Leaflets which have been widely acclaimed as a breakthrough in gardening advice.

The chapter headings give a clear indication of the subjects covered. It will be apparent that this book is not intended to be a comprehensive guide to vegetable growing, but each subject included is dealt with in depth and at the end of each chapter the conclusions are summarized in tabular form for easy reference.

The Contributors

The Editors, who are also contributors, are:

Professor J. K. A. Bleasdale, *Director,*
National Vegetable Research Station,
Wellesbourne, Warwickshire

Dr. P. J. Salter,
Head of the Plant Physiology Section,
National Vegetable Research Station

The other contributors are also members of the staff of the National Vegetable Research Station. They are:

Dr. R. T. Burchill,
Head of the Plant Pathology Section

T. J. Cleaver
Head of the Field and Glasshouse Section,
formerly of the Soil Science Section

Dr. D. Gray,
Plant Physiology Section

G. A. Wheatley,
Head of the Entomology Section

Contents

1 Space to grow

Most gardeners are restricted in the area they can devote to vegetables so spacing crops to get maximum yield makes sense. The allotment holder is more fortunate, but even here getting the most from each square yard is a desirable objective because nearly all the chores and costs of vegetable growing are entirely dependent upon the area cultivated. Why dig, fertilize, weed, and water more land than is necessary to produce an adequate supply?

Most gardening books will tell you precisely the distance apart for, say, rows of onions and precisely how far apart along the row to thin the plants once they have emerged. Few of them seem to agree with each other and even fewer explain why they think their spacing is right. But in general they all tend to be over-generous in the space given to all crops. If you want to get the best return for your efforts, research has now produced results that can help you decide the best spacing for your crops. What gardener could test fifty different spacings for a runner bean crop to find the best? Research *has* done this, and although it has been directed towards commercial requirements the results are equally useful to the gardener and they are discussed later in this chapter. Research has also established the principles of vegetable-crop spacing and has shown how to control not only the yield per unit area but also the size of the individual bulbs, roots, or hearts. For example, spacing alone can determine whether you produce a crop of small pickling onions or more normal-sized ones for casseroles and frying. Indeed, spacing is such a potent tool that it can produce new crops like leaf-lettuce and mini-cauliflower, both of which are described later.

All gardening is fundamentally an attempt to control the conditions under which chosen plants grow. Our control of most of the things that matter is at the best only partial—we rarely get rid of all the weeds for all of the time, or have exactly the perfect balance of soil conditions. The beauty of spacing is that we really can have precise control over it. The difficulty is that we need to know what the right spacing is before we can exercise that control. Let us start resolving the difficulty by explaining some of the principles involved.

POPULATIONS AND PATTERNS

The population of plants is the number per unit area and the pattern, obviously enough, describes the way in which the population is arranged. Both are of vital importance and affect each other. Common sense will tell us that this is so. For example, it is obvious if we have only one cabbage plant per acre that the pattern does not matter. As soon as we have two cabbages per acre we can see that if we planted both of them in the same hole they would not do as well as if they were well separated. We can also see that even if the cabbages grow to a size warranting an entry in the record books the yield per acre would be pathetic. If we gradually increased the number of cabbages on our acre and kept them as evenly arranged as we could, the yield would at first increase completely in step with the increase in the number of cabbages. Each cabbage would be more or less the same size and that size would be the maximum that the conditions would allow. If we gradually increase the number still further, but kept them evenly arranged, each cabbage would be smaller because it would gradually have increasing competition from its neighbours. However, the yield per acre would be greater even though the steps in the increase would no longer be as great for each extra cabbage as they were when there were only a few cabbages. As numbers increase even more the individual cabbages get even smaller and eventually a point is reached where total yield per acre is at a maximum.

This type of response to increasing plant population happens with all crops and we know from research the populations needed for the maximum yields of most vegetables. However, it is at precisely this point that vegetables begin to differ from each other. Some crops, such as carrots, hold their maximum yield as populations are further increased. This is very convenient as it means we can have maximum yield and can control the size of the individual plants. For example, commercial growers can produce similar yields per acre of medium-sized carrots for shop sales, or of very small carrots for canning whole, simply by using higher populations for the canning crop. However, some other vegetable crops are not so obliging for, although the total amount of vegetation produced per acre will remain at a maximum as population is increased, the yield of the part we want may begin to decline. With red-beet this decline is quite rapid, with other crops the decline still occurs but is so slight that we can accept it provided increasing the population gives us the control of size we want. An example of this is onions, where the highest populations used to produce small bulbs for pickling give slightly less than maximum yield, but if you want pickling onions this doesn't matter.

Unfortunately, it is also true that the populations giving maximum yield of some vegetable crops give individuals that are too small for normal use. Of course, people differ in the size they would consider normal or acceptable. It is precisely at this point that disagreement between spacings recommended in different gardening books arises and even more acutely between such spacing and those found by research and commercial growers to give the best yields. Most of the professional gardeners who write gardening books have decided, through the influence of shows and a natural tendency, that large is beautiful. Most commercial growers have found that the housewife thinks small is beautiful—although there are obviously limits in both directions. Onions again provide a good example in that at the populations giving maximum yield per unit area the individual bulbs are that awkward in-between size of about $1\frac{1}{4}$ in. (32 mm.) in diameter—slightly too big for pickling and needing too much peeling to be practical for stews or frying. Consequently, commercial growers use a population somewhat below that giving maximum yield in order to get a better size—mostly above $1\frac{1}{2}$ in. (38 mm.) in diameter. Their loss in total yield is minimal, but if you want to take pride in producing large onions you can only do this by losing even more yield. But let us now turn to the pattern of plant arrangement and again let common sense be our guide.

The soil containing the nutrients and water our crops need is evenly distributed. The solar energy our plants are to convert into food for us is also evenly distributed, so it seems to be reasonable to expect our plants to be able to make the maximum use of these resources if they, too, are evenly distributed. Why then do we grow many crops in rows with the plants crowded within the row (Fig. 1.1) and a relatively large space between the rows? If what our common sense tells us is right, then each plant ought to be in the middle of a circle representing its space and the pattern should be one giving the minimum overlap of these circles (Fig. 1.2) with the maximum amount of the ground

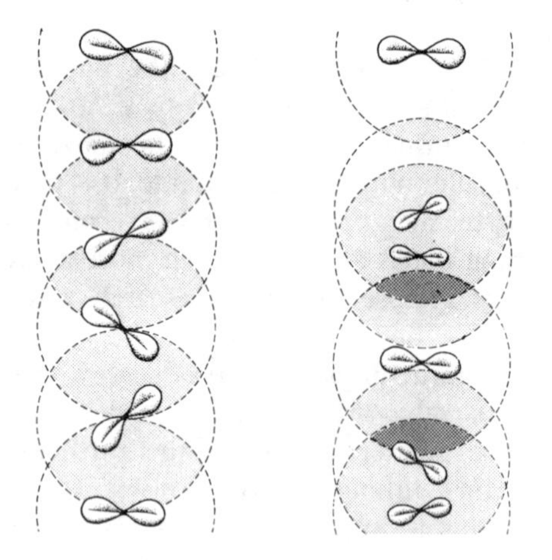

Fig. 1.1. Regularly spaced plants in a row (left) all suffer equally from competition represented by the shaded areas where the circles of each plant's domain overlap. When irregularly spaced the competition is also irregular and so the plants become more uneven in size. In both rows the plants are competing, but there is space between the rows not being used.

covered. Why don't we do this? Is our common sense nonsense? Research has shown that, for all intents and purposes, common sense is right so the mystery deepens.

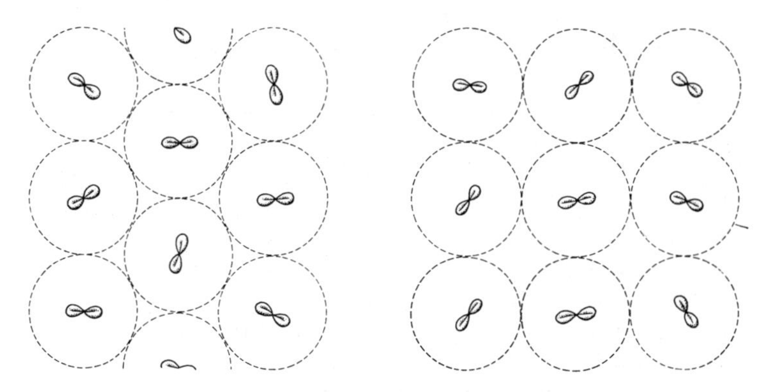

Fig. 1.2. The left-hand arrangement of plants makes fullest use of the land and reduces competition between plants to a minimum. A square arrangement (right) is less effective.

To find the answer we have to go back to those school history lessons and recall Jethro Tull and Turnip Townsend from the recesses of our minds. They were agriculturalists concerned about the need to fallow land every few years to reduce the weed problem. Jethro Tull invented a seed-drill, the first, so that crops could be sown in rows rather than broadcast. He also invented a hoe pulled by a horse to travel between the rows to kill the weeds. He and others did experiments with rows of wheat 6 ft. (183 cm.) apart and compared the yield with that obtained from broadcast crops. He got better yields from the rows but, we now realize, only because he was controlling devastating weed-growth by horse-hoeing. Instead of leaving the land fallow, turnips were grown in rows for feeding to sheep and agriculture in the eighteenth and nineteenth centuries gradually moved over to row-cropping. Late in the eighteenth century vegetables were grown in market gardens on the edge of towns and the system was to grow crops in strips about 4 ft. (122 cm.) wide separated one from the other by narrow paths. The seed for crops was broadcast on these strips and women and children were employed to remove the weeds and to thin the plants to get an even distribution over the strip. The whole strip or bed could be reached from the paths.

The industrial revolution caused the towns to expand, destroying the traditional market gardens and diverting the labour for weeding into factories. The railways enabled vegetables to be brought from further afield and all these factors combined to induce the row-cropping farmers to grow vegetables. They did it in the only way they knew and so row-cropping became established as the method of vegetable-crop production. The rows were very wide apart because the horse-hoe was still the prime method of weed control.

Research has brought us full circle. In the eighteenth century the cheapness of labour made it possible to hand-weed evenly distributed vegetable crops grown in beds. In the twentieth century herbicides control the weeds and beds are again used, the 'paths' being the wheel tracks of a straddling tractor. The modern patterns of crop arrangement are as near-even within the bed as it is practicable to achieve.

Most of the potent selective-herbicides available to commercial growers are not on sale to gardeners, although some are available which can be used on certain crops. Generally, the gardener will rely on hoeing or hand-weeding to control weeds so it is sometimes sensible to adjust the pattern of plant arrangement to some extent and use close rows to make weed control easier. Let us look in more detail at the weed problem.

COMPETITION FROM WEEDS

If we grow crops of red-beet and onions in rows 18 in. (45 cm.) apart we can hoe between the rows and leave an undisturbed band of soil coincident with the row. Thus, between the rows will be weed-free, but within the row there will be a band of weeds. On other plots we grow the crops in an identical manner, but arrange things so that the band of undisturbed soil with the weeds is exactly the same width but is between the rows of crop. Thus, the weeds are the same in both lots of plots, they only differ in their position. On still other plots we assiduously remove all the weeds with the minimum of disturbance to the crop and whilst the weeds are very small. When we harvest the crops we find that the onions with weeds are a write-off regardless of whether the weeds were in the row or between the rows. In experiments with red-beet the relative yields for such a comparison were: weed-free 100 per cent, weeds in-the-row 78 per cent, weeds between-the-rows 64 per cent.

We learn from this that the broad leaves of red-beet can suppress weeds but that the slender tubular leaves of onions have no smothering capacity. We also find, somewhat surprisingly, that weeds nearer to a smothering crop are less damaging than those further away. This is because the nearer weeds are always under some suppressing influence, whereas those further away and between the rows had a non-suppressed start, got above the crop, and so eventually exerted a big effect on the crop rows.

Research of this kind has shown us how to use the crops themselves to help us to control the weeds. For example, if an evenly distributed crop of carrots is kept clean until it has two true-leaves (not the seed leaves but the 'true' or feathery ones), then the crop itself will smother all subsequent growth of annual weeds. Until we had persistent herbicides for commercial carrot crops the biggest problem was not weeds in the beds but weeds growing in the tractor wheelings where there was no competition from the crop.

Generalizing we can say that crops with broad smothering-type leaves can be used to suppress weeds. The essentials are as rapid and complete a shading of the soil surface as is practicable together with diligent weeding up to as near to that time as is possible.

From what has been said previously it follows that as even a distribution of the crop plants as is possible will give the quickest complete smothering of weeds and the quickest and most complete usage of all the sunlight available to the crop. However, this may only be marginally quicker than can be obtained by using fairly close rows and these do have the advantage of allowing us to remove most of the weeds by careful hoeing with a small hoe.

Onions are not going to help us to smother weeds so we may choose to lose some yield by using wider row spacings to make hoeing easier. But even here the closer the rows the better the yield.

Before going on to consider the spacing of some of the most popular vegetable crops we can summarize the principles that will guide us.

The *population* is the number of plants per unit area; ideally we will want to use the population or populations giving maximum yield, but if the individual's size is too small we may have to compromise and use a lower population. With some crops a wide range of populations give maximum yield and with these we can use population to control size without loss of yield, higher populations giving us the smaller sizes.

The *pattern* in which the plants are arranged should be as near even as possible as this ensures maximum yield and will generally help by suppressing weeds. However, some compromise is desirable to make hoeing easier, particularly if the slight unevenness of distribution has only a minimal effect on yield and the crop is one which does not have smothering foliage.

It also needs to be stated that in the later detailing of the spacing for crops, the assumption is made that the soil is fertile and that water will not be seriously limiting. Desert farming is practised successfully in many arid parts of the world and its major characteristic is the sparseness of the planting. To survive every plant must have a large area for the roots to obtain enough water (see Chapter 4). We are not interested in such extremes, although just occasionally our temperate climate can seem more like a desert. Similarly, neither are we catering for situations where the nutrient supply is so deficient that even widely spaced cabbages would give hearts no bigger than tennis balls. However, even in these extreme situations the principles governing the patterns of plant arrangement still apply. An even pattern of plant arrangement ensures the full exploitation of whatever resources of soil and situation exist.

TRANSPLANTED CROPS

Cabbage

In experiments with summer cabbage, spacing the plants 14 × 14 in. (35 × 35 cm.) gave the highest yield of small heads which were big enough for most families. Wider spacing up to 18 × 18 in. (45 × 45 cm.) did not reduce yield but gave bigger heads and slightly earlier hearting. Although this difference in spacing may seem small the closer spacing gives 65 per cent more plants than the wider one, emphasizing the need for a measuring stick when planting. It may be sensible to grow some cabbage at each of the two spacings to meet the varied demand that can exist even within a family. Few research results are available for winter or spring cabbage. With the latter, rows 12 in. (30 cm.) apart with plants 4 in. (10 cm.) apart seems likely to be effective if plants are removed for use to give a final spacing of 12 × 12 in. (30 × 30 cm.) for hearted greens. Winter cabbage are probably little different in their space requirements to summer cabbage.

Cauliflower

A range of cauliflower types needs to be grown if the gardener is to produce throughout the year. As a general guide the later the crops are planted in the year the wider the spacing required. For the early summer cauliflower planted from pots in March, 21 × 21 in. (53 × 53 cm.) will be an adequate spacing. For late autumn-maturing types and for winter cauliflowers 27 × 27 in. (68 × 68 cm.) or even 30 × 30 in. (76 × 76 cm.) will be necessary to produce a large curd because of the mass of foliage produced by these types. With summer and autumn cauliflower, experiments have shown that the spacing for maximum yield is very dependent on water supply. If you are able to water this crop frequently, spacings from as close as 17 × 17 in. (43 × 43 cm.) will give similar total yields to wider spacings, but the closer spacing will naturally give small curds. In drier conditions, spacings from 24×24 in. (61×61 cm.) to 34×34 in. (86×86 cm.) are more suitable (see Chapter 4).

Mini-cauliflower

If certain varieties of cauliflower are grown at very close spacings they produce high yields of miniature curds ($1\frac{1}{2}$–$3\frac{1}{2}$ in. (38–89 mm.) diameter) each suitable as a portion for one person. They are ideal for home freezing, particularly as another consequence of the close spacing is that all the plants tend to produce curds which are ready for harvest within a few days of each other.

A suitable spacing is for the rows to be 9 in. (23 cm.) apart with the plants 4 in. (10 cm.) apart within the row. Ideally, this should be achieved by spaced-sowing rather than transplanting as this gives a more uniformly maturing crop. Several varieties, although not all, of the normal types of early summer cauliflower are suitable, notably 'No. 110', 'Garant', and 'Predominant'. Some seedsmen also offer special varieties. Several sowings can be made within a year to provide a succession.

Calabrese

Plant populations ranging from 0·5 to 10 plants per square foot (5 to 107 per square metre) all give similar yields, but quality tends to be better at higher populations. Two plants per square foot (21 per square metre) will give good-quality spears with about half the yield from terminal shoots and half from side-shoots. The plants are remarkably insensitive to the pattern of arrangement up to a between-row distance of 24 in. (61 cm.). Thus, 3×24 in. (8×61 cm.), or 4×18 in. (10×45 cm.), or 6×12 in. (15×30 cm.), would all be expected to yield equally well.

Wider spacings will give larger spears with more of the yield coming from side-shoots. Very close spacings suppress side-shoots and give smaller terminal spears which tend to be all ready for harvest at the same time. This can be useful if you plan to deep-freeze some for later use.

Brussels sprouts

Most gardeners treat sprouts as a 'cut-and-come-again' crop, picking off each plant the sprouts as they become ready. If this is what is wanted then a spacing of 36×36 in. (91×91 cm.) is as productive as closer spacings, especially if the aim is to allow the individual sprouts to reach full size before they are picked. This wide spacing is also desirable to allow access for picking.

An alternative approach is to adopt the method used by commercial growers to produce sprouts for freezing. This involves a closer spacing of 20×20 in. (50×50 cm.), removal of the top growing-point and small leaves of the plant when the biggest sprouts are about the size of your little fingernail, and a single harvest taken when the very bottom sprouts are just past their best. This will give you a bulk supply of small sprouts suitable for home freezing and it can be a good tactic where pigeons are expected to eat most of the standing crop in the winter.

Lettuce

The normal 'butterhead' type of lettuce grown during the summer months is adequately spaced at 12×12 in. (30×30 cm.). Lettuce are almost the perfectly circular crop, corresponding very closely to Fig. 1.2. Some economy of space can be made by a triangular rather than a square pattern of planting. Circles of diameter 12 in. fit into this pattern if the plants are 12 in. (30 cm.) apart in rows 10½ in. (27 cm.) apart. This pattern of planting is particularly valuable for frame or cloche crops which can be grown as close as 9×8 in. (23×20 cm.) on a triangular pattern.

Leaf lettuce

This term is generally used to describe lettuce of the 'Salad Bowl' type that never form true hearts, but you can obtain a supply of leaves from normal varieties. These are transformed by growing them at such close spacing that they do not heart and all they produce are leaves. The spacings are so close that the crop is not transplanted, but it is included here with conventional transplanted lettuce for continuity and completeness.

Research has shown that the most suitable varieties are certain cos types, notably 'Paris White Cos', 'Lobjoits Cos', and 'Valmaine'. Other types and varieties can be bitter at the early stage of growth at which this crop is harvested. Seed is sown in rows 5 in. (13 cm.) apart (if you adopt a bed-system it is easier to sow rows across the bed) with the aim of getting 12–15 plants per foot (30 cm.) of row. In mid-season the crop is ready for harvest in 40 days from sowing as compared with 60 days for conventional crops. Early in the season the corresponding figures are 50 and 80 days. After cutting the stumps will regrow to produce a second crop. The equivalent of four or five normal hearted lettuce per week from mid-May to mid-October can be obtained from as little as 5–6 square yards (4–5 square metres) of garden by following the sowing sequence shown in Fig. 1.3. Slightly less than one square yard should be sown on each of ten dates. For the later sowings ground cleared from earlier sowings can be re-used.

Freshly harvested, these leaves are very enjoyable although those who like the crisp heart of conventional lettuce may be disappointed.

Celery

No spacing experiments have been carried out with trench celery as with this crop the area needed for the trench dominates the crop spacing. Self-blanching celery, the sort grown above ground without earthing-up, has been studied and to produce sticks you will need approximately 120 square in. (760 square cm.) per plant. This can be 12 × 10 in. (30 × 25 cm.) or even 13 × 9 in. (33 × 23 cm.) (near enough!) or, best of all, 11 × 11 in. (28 × 28 cm.). How can we be so precise? If the spacing is wider than this, blanching of the sticks does not occur. Indeed, to get good self-blanching at this spacing you will need high soil-fertility and plenty of water. But if you get everything right so that you obtain rapid growth to give you succulent celery, then it will be a good size and well blanched at 120 square in. per plant. One can almost say that if you can't produce the conditions that

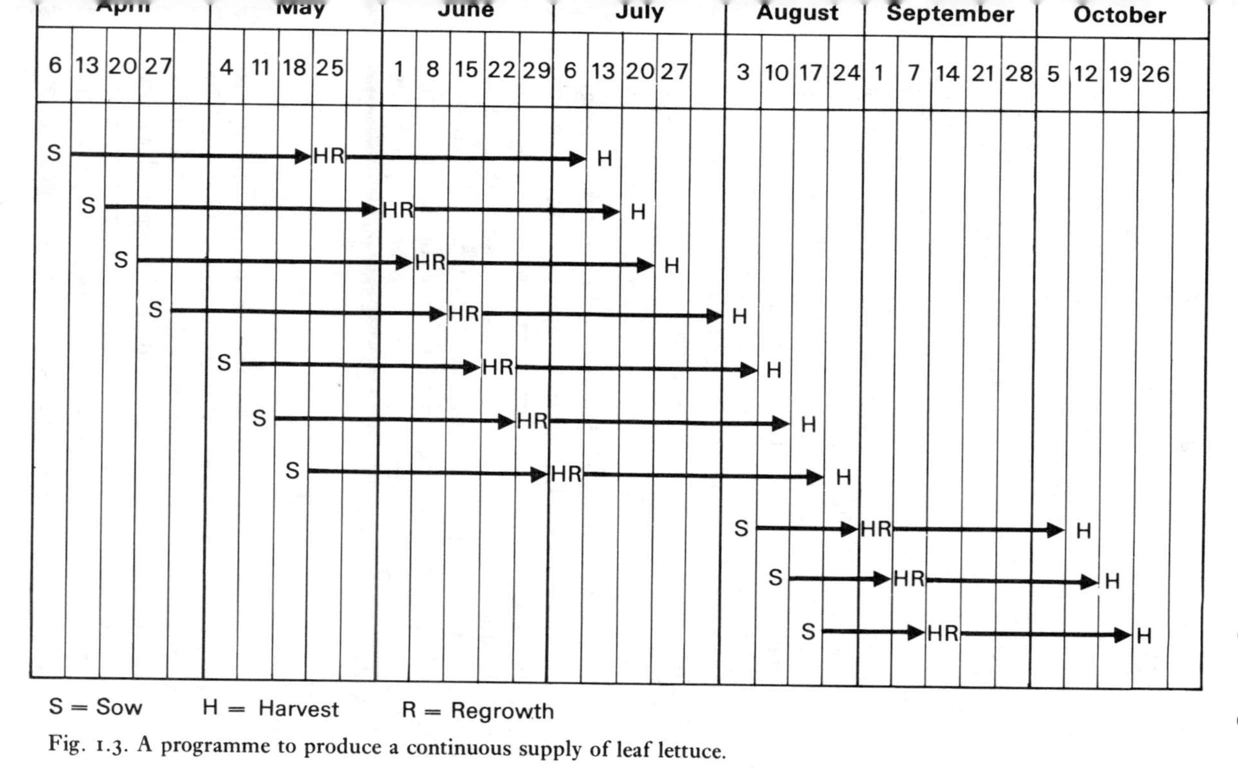

Fig. 1.3. A programme to produce a continuous supply of leaf lettuce.

make this spacing precisely right, then you won't be able to grow good celery.

Of course, you can plant closer and you will get higher yields but of much smaller sticks. A spacing of 6×6 in. (15×15 cm.) gives slender hearts which are very good either cooked or raw. However, if you are buying the plants such close spacing can be expensive. You can always compromise at 9×9 in. (23×23 cm.)!

Leeks

A spacing of 12×6 in. (30×15 cm.) gives maximum yield and normal-size leeks. The 12 in. spacing between rows gives enough room to plant into a shallow V-shaped trench and enough room to invert the V by earthing up as the plants grow. The bigger transplants will give you bigger leeks at harvest.

Closer spacing can be used without loss of yield to produce more slender leeks if these are preferred.

Onion sets

As a sensible economy always buy the smaller sets even though they cost more per pound. They are cheaper per plant and much less liable to bolt. If you are after the maximum yield per unit area of ground, then rows 10 in. (25 cm.) apart with the sets 2 in. (5 cm.) apart along the rows will give high yields of medium-sized bulbs. Wider spacings along the row up to, say, 4 in. (10 cm.) can be used if land is not precious and larger bulbs are wanted. Any wider spacing than this and yield is severely reduced.

Potatoes

Seed potatoes are not seed in the true sense. Like onion sets they are dormant transplants. We can, however, liken the bud or 'eye' on a seed potato to the individual seed of such plants as lettuce. In the same way that not all the lettuce seed we sow will produce plants, so not all the eyes grow to give us potato plants. However, from any one seed potato, unless it is very small, more than one eye will grow so we essentially get a clump of two or three or even more potato plants (Fig. 1.4). From what has been said about the effects of the pattern of plant arrangement you will already be alarmed at the prospects of such an uneven pattern of arrangement arising from the clumping.

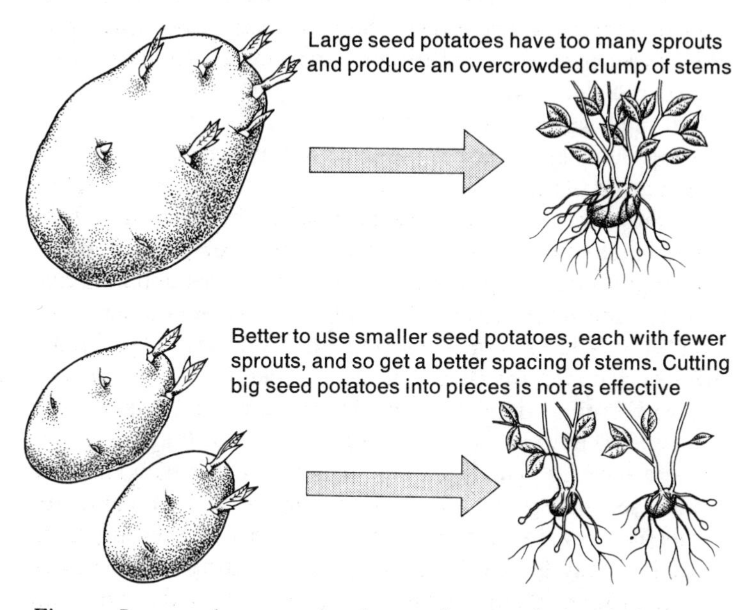

Fig. 1.4. Large seed potatoes give clumps of stems reducing yield.

The effects of this clumping are twofold. It reduces yield somewhat, and it means that a lot of potatoes are crowded together in the soil as the new crop grows. It can be shown that this produces such congestion that some tubers are forced to the surface—they don't grow there, they are forced there by their own growth and that of other tubers.

We can reduce clumping and this overcrowding by using small seed potatoes. If we used very small seed each weighing about $\frac{1}{3}$ oz. (10 g.) they will not produce many stems or many tubers each. But if they are planted evenly and moderately closely spaced they will produce a high yield, they will not need any earthing-up at all, and the whole crop can be grown on-the-flat by planting the small seed 4 in. (10 cm.) deep.

Why should moderately close spacing be essential? The answer to this lies in the fact that potato plants behave exactly like all the other plants we've discussed so far. The more space you give them the bigger they get—or, in the case of potatoes, the more space you give a clump the more tubers and the bigger tubers you will harvest per clump. Thus, wide spacing will, even when small seed is used, lead to an overcrowding of the tubers in the ground and some will be forced to the surface and become green in the light.

So far we've arrived at two reasons for using small seed: first, to reduce clumping, and second, to reduce greening. Both these objectives could be achieved by cutting large seed into planting pieces, but this should only be done if you cannot obtain small seed. Those of you who are mathematically inclined will see why, once it is appreciated that the number of eyes on a potato is directly proportional to its surface area and not to its weight. Thus, if we have two tubers one weighing 1 oz. (28 g.) and the other 2 oz. (56 g.) the larger one will have fewer than twice as many eyes as the smaller one. Weight for weight, and remember seed potatoes are sold by weight, we get more of what we want, eyes, if we buy small seed. No amount of cutting of larger seed can produce the same effect.

Farmers generally plant about 1 ton of potato seed per acre (2·5 tonnes per hectare) for maincrops. The results of the research outlined above are making them realize that this is too inexact and that the planting rate must take account of the seed size. Experimental results have shown that as little as 6 cwt. per acre (0·75 tonne per hectare) of small seed can produce a crop virtually identical to one from 2 tons per acre (5 tonnes per hectare) of large seed.

The farmer has available to him tables of planting rates for different sizes of seed potatoes for each of many different varieties produced by the Agricultural Development and Advisory Service. The gardener can work things out for himself or use Table 1.1 as a guide. For maincrop plantings, the aim should

Table 1.1

A guide to the within-row spacing of potatoes as determined by the size of seed when grown in 30 in. (76 cm.) rows

Seed size:	Large	Medium	Small	Very small
Number per 14 lb.:	under 75	75–100	100–125	over 125
(Number per 10 kg.):	(under 120)	(120–160)	(160–200)	(over 200)
'King Edward'	17 in. (43 cm.)	15 (38)	14 (36)	12 (30)
'Maris Piper'	17 (43)	15 (38)	12 (30)	10 (25)
'Pentland Dell'	16 (41)	12 (30)	11 (28)	9 (23)
'Pentland Crown'	12 (30)	11 (28)	10 (25)	8 (20)
'Pentland Ivory'	11 (28)	10 (25)	9 (23)	8 (20)
'Majestic'	11 (28)	10 (25)	8 (20)	8 (20)
'Desirée'	11 (28)	10 (25)	8 (20)	7 (18)

be to plant 8 eyes per square foot (86 per square metre) of cropped area using whole seed. This will give 2–3 'plants' per square foot (21–32 per square metre).

Early potato crops introduce another dimension and that is the importance of the food reserves in the seed potato for the early growth and yield of the crop. These reserves are also important as an aid to recovery from frost. If the early growth is cut-back by frost it is vital that there are adequate reserves for re-growth. The dilemma of wanting a minimum of clumping and a maximum of reserves is resolved by using relatively large seed-tubers which have been stored in conditions that give only one or two sprouts from the 'rose' end of each seed (Fig. 1.5). To do this you have to buy the seed potatoes as early as possible—preferably in September or October—and keep them somewhere warm, about 65°F (18°C), until you can see the eyes at the rose-end breaking into growth. The tubers should then be stored cool, around 45°F (7°C), until near planting time when warmth and light are used to forward the sprout. The early sprouting of the rose-end eye suppresses the sprouting of the other eyes on the seed-tuber and so gives us what we need for early crops—large reserves and few sprouts.

Of course, it won't be as good if you have many sprouts and reduce them to only one or two by rubbing the others off at planting. If you do this this you will have already dissipated some of the reserves by producing unwanted sprouts. Still, it is better than planting with a lot of sprouts on the seed-tuber.

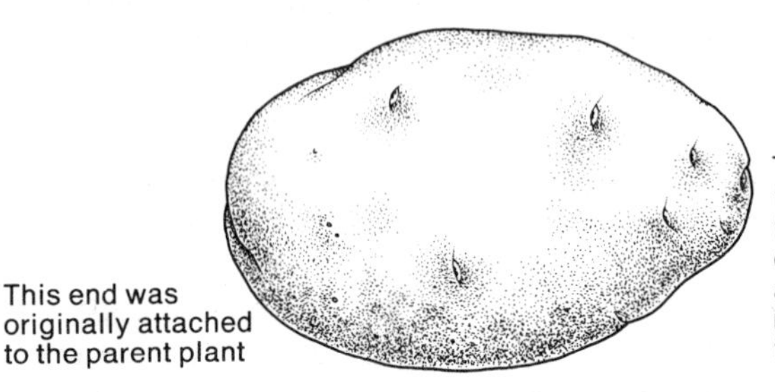

Fig. 1.5. The eyes at the 'rose' end of seed potatoes can be started into growth soon after harvest. If they are then held in cool conditions until put to sprout in the spring, only one or two eyes will produce shoots. This is what is needed for early production. If all sprouting is delayed till the spring more eyes will sprout and this is what is needed when small seed potatoes are used for maincrops.

Having got large but sparsely sprouted seed it can be planted to give much the same *plant* spacing as for maincrops. Namely, 2–3 main-stems per square foot (21–32 per square metre) as counted shortly after the stems come up through the soil. This will mean planting 3–4 sprouts per square foot (32–43 per square metre) as not all sprouts manage to emerge.

Incidentally, with the maincrop seed we want to get as many eyes to grow as possible so the pre-planting treatment is different from that for early seed. Maincrop seed should be stored cool, between 45 and 50°F (7 and 10°C) until early spring. You will then find many of the lower eyes as well as those at the rose-end will break into growth.

If you try growing potatoes on the flat you can be really lazy and plant them as you dig. If when they are just through you cover the shoot with plant pots, then a quick spray with a paraquat/diquat mixture (Weedol) will see off the young weeds and by the time the next crop of weeds comes through you will have a good smothering canopy of potato leaves. Of course, you must not forget the other advice in this book about the nutrient and water needs of the crop, and this 'lazy' method of culture is less good on heavy soils. These are best dug in the autumn as then the winter's frosts can help you to produce a good tilth.

Shallots

Each shallot planted produces a clump of plants. To avoid having too many plants per clump it is better to use sets each weighing about ⅓oz. (10g.) (i.e. about 50 per lb. or 100 per kilo). These will give the maximum yield of good-sized shallots given 48 square in. (300 square cm.) of space each. This can conveniently be 6×8 in. (15×20 cm.) or 4×12 in. (10×30 cm.).

If larger sets are used, say of a size giving about 20 per lb. (45 per kilo), then it is better to use a wider spacing of about 6×12 in. (15×30 cm.), otherwise bulb-size at harvest will be reduced. The total yield will be similar to that obtained with the smaller sets which are, therefore, always to be preferred, especially as they are also less prone to bolting.

Tomatoes

Detailed experiments have shown that when evenly distributed 'on-the-square' outdoor tomato plants each need about 2½ square ft. (approximately ¼ square metre). One of the main advantages of these close spacings is the increase in early yield that is obtained. For example, 5 square ft. per plant gave only half the early yield of plants with 2½ square ft. each, but there was relatively little difference in total yield at the end of the season when a larger-growing, more vigorous variety was used.

Thus, as a general guide, outdoor bush-types such as 'Amateur' or 'Sleaford Abundance' should be planted 19×19 in. (48×48 cm.). If planted in 4 ft. (122 cm.) wide beds, then two rows per bed spaced 15 in. (38 cm.) from the paths and with the plants 15 in. apart gives convenient arrangement of the required population. This spacing is also suitable for staked varieties, such as 'Moneymaker'.

Plants can be expensive, so it is tempting to use wide spacing. However, it must be remembered that your early yield, obtained when tomatoes are relatively expensive, will be greater with the closer planting and your end-of-season glut correspondingly less. If you have a cold frame you can, with the new techniques described in Chapter 2, easily grow your own transplants.

CROPS GROWN FROM SEED

Most of the spacings for these crops will involve using rows closer than those you may consider normal. If you use a bed-system as described earlier you will find it easier to use a small hoe between the close rows if you run the rows across the beds rather than along them. A commercial grower runs his rows along the beds which are straddled by his tractor, but this is the only way he can do it. He would also use a tractor-drawn hoe if for any reason his herbicides failed to work. So don't be put off if a 'professional' tells you you've got your rows running the wrong way.

Carrots

This is the crop on which many of the principles of crop spacing already described were first established. The more recent research has revealed another important and probably general principle. Simply stated, this reveals that *allocating* each plant an even amount of circular space is not enough to ensure an even-sized crop of roots. This is because those plants which emerge first remain ahead of others and, as it were, 'take' more space than they are entitled to. Consequently, they are bigger than they should be on the basis of the area they occupy, and their neighbours are smaller.

Thus, it has become important to get a rapid and even emergence of the seed in order to improve crop uniformity (see Chapter 2). Precise mathematical relationships describing the effects of spacing and the time-spread of emergence have been developed and are being used on behalf of commercial growers who, for example, aim to produce high proportions of their crops in the limited size-ranges required for canning whole or for pre-packing.

In the garden such sophistication is not warranted, but it is important to realize that if you are growing populations which will give maximum yield, quite a *number* of the plants (sometimes as many as 20 per cent) will be too small to use. Their total weight will be small and all the spacings referred to in this chapter take account of this and are those which give maximum usable yield even though some of the total yield will be unusable. Gardeners seem to abhor this seeming waste and this is one of the reasons why gardening books tend to recommend wider spacings than those described in this chapter. These wider spacings give few or no waste plants, but they also give lower usable yields. Commercial growers have no such illusions.

Carrots can be sown from February to July and can be harvested as soon as the roots are big enough to eat. Thus, if the objective is to produce medium-sized carrots they can always be obtained by allowing sufficient time to reach that size and then harvesting them. If the time available is short, as when early crops are wanted or when the land cleared from early potatoes is re-used, then competition between the plants must be minimal to get the most rapid growth of the individual plants. In these situations low populations, 5–7 plants per square foot (54–75 per square metre) are used. The yields obtained will be low, but this is acceptable in view of the short-term nature of the crop.

Higher yields are obtained by allowing a longer period for growth, and for maximum yields February/March sowings are harvested from November onwards. However, such long-term crops are generally of poorer quality, particularly in their shape and skin, than crops sown in early May for harvest from November onwards. For such crops populations of 15–20 plants per square foot (160–215 per square metre) will give the highest yields of medium-sized carrots.

In the garden both short-term and long-term crops should be sown in rows no more than 6 in. (15 cm.) apart. For a reminder on how to use the crop to help you to control the weeds, I refer you back to the part of this chapter on 'Competition from weeds'.

The best place to store carrots for winter and spring use is in the ground where they are grown. They must be protected from frost and this can be done by piling autumn leaves on to them. It is as well to put some slug bait down under the leaves and to be sure to have a settled depth of at least 9 in. (23 cm.) of leaves. Straw 1 ft. (30 cm.) deep is an effective alternative.

Parsnips

The maximum yield of roots of the larger-rooted varieties such as 'Offenham' is likely to be obtained with 3 plants per square foot (32 per square metre). Somewhat higher populations will give slightly lower total yields. With smaller-rooted varieties, such as the canker-resistant 'Avonresister', the maximum yield of roots is obtained at 6–7 plants per square foot (64–75 per square metre) but at this population the roots will be small at around 1½–2 in. (38–50 mm.) crown diameter. This size is becoming more favoured by housewives and can also be obtained with the larger-rooted varieties by growing them at 6–7 plants per square foot. If you want most of the roots to be more than 2 in. in crown diameter the population should not exceed 2 plants per square foot (21 per square metre). It should be noted that 'Avonresister' should only be grown at this population if you expect to be troubled with canker. At this lower population it will yield 25 per cent less than the larger-rooted varieties because of its inherently smaller root-size.

Experimental results indicate that provided the ratio of the between-row spacing to the within-row spacing (the rectangularity) does not exceed 2½ : 1, yield is not reduced. Thus, for 2 plants per square foot (21 per square metre) the maximum row spacing would be 13·4 in. (34 cm.) with the plants 5·5 in. (13·5 cm.) apart in the row—12 × 6 in. (30 × 15 cm.) giving a rectangularity of two may seem simpler. For 6 plants per square foot (64 per square metre) rows 7½ in. (19 cm.) apart with the plants 3 in. (8 cm.) apart would be suitable.

Red-beet

With early crops of the 'Detroit' type of red-beet the objective is to minimize competition as much as is consistent with getting a reasonable yield. This is achieved by establishing 5 plants per square foot (54 per square metre) in rows no wider than 7 in. (18 cm.) apart. The within-row spacing can easily be calculated because in each square foot there will be 144÷7=20·6 in. of row. There are to be 5 plants in this length of row so they must be 20·6÷5=4·1 in. apart, so we would aim at a 4 in. spacing in rows 7 in. apart for early beet. For the longer-term maincrop up to 15 and not less than 10 plants per square foot (161 to 107 per square metre) will give the maximum yield of medium-sized roots if the rows are no wider than 12 in. (30 cm.) apart. If you want to produce small beet for pickling whole, don't be tempted to use more than 20 plants per square foot (215 per square metre). Red-beet are one of those crops where total yield is very much depressed at high populations and even at 20 plants per square foot you will lose some yield.

Onions

We have already dealt with onions from sets under 'Transplanted crops'. Onions from seed may be sown in the early spring or in August provided these latter are of the new over-wintering varieties. The sowing date is critical for the over-wintered crop and rapid emergence is also vital. Crops in the north of Britain must be sown no later than the first week in August, whereas those in the south should be delayed until the middle of the month. The reason this is so critical is that if the plants are too small when the winter starts many will be killed and if they are too big many of them will go to seed rather than form bulbs. We've already learnt that spacing also affects plant size so it will be evident that overcrowding of the August-sown crop could reduce plant size and so reduce the chances of winter survival.

The aim should be to have in the spring 6–8 plants per square foot (65–86 per square metre) in rows no more than 12 in. (30 cm.) apart. To allow for some winter losses it is as well to establish 10 plants per square foot (108 per square metre) in the autumn, using any spring thinnings for salads.

The best population for the spring-sown crop is 8 plants per square foot (86 per square metre), again the rows not being wider than 12 in. (30 cm.), and there are some indications that 9 in. (23 cm.) is better. This population will give bulbs mostly more than $1\frac{1}{2}$ in. (38 mm.) in diameter, but there will be a few smaller ones of pickling size. The population giving the maximum yield of onions varies somewhat depending upon the variety but is about 20 plants per square foot (215 per square metre). However, the onions would be too small for normal purposes and somewhat too big for pickling. If you want to grow small onions there is little loss in total yield if the population is increased to 35 plants per square foot (376 per square metre) and this will give you a much better size for pickling.

Salad onions

Commercially the variety 'White Lisbon' and selections from it are most widely used for the salad crop. Using this variety, experiments have shown that both for the overwintered and summer crops 30 plants per square foot (323 per square metre) in rows 4 in. (10 cm.) apart or in 3 in. (8 cm.) wide bands with 12 in. (30 cm.) between centres gives the best yields. Sowing a band rather than a single row has also been applied commercially to other crops, notably carrots. The aim is to reduce the effects of overcrowding within the row by spreading the row from a single line to a broader strip. It works well, but it can increase the chore of weeding when, as is usually the case with gardeners, herbicides are not used.

French beans

All the research on the spacing of this crop has been concerned with commercial crops grown for freezing. Such crops are grown to be harvested by machine on one occasion only. In the garden, pods are picked when they are ready and several harvests are made over a period. It seems probable that this difference in harvesting technique will affect the spacing that should be used.

The research for commercial crops has shown that 3–4 plants per square foot (32–43 per square metre) can be expected to give the maximum yield and that an even pattern of plant arrangement is an advantage. Access for hand-picking will, however, tend to dictate using rows about 18 in. (45 cm.) apart. Thus it would seem that the plants should be 2 in. (5 cm.) apart in the row. Certainly they should be no closer than this and it seems that if they are more than 4 in. (10 cm.) apart yield will be lost.

Runner beans

This is one of the most popular garden crops and research results are available to guide us on its best spacing. In these experiments the crop was supported on canes arranged as in Fig. 1.6, and the following combinations of path width and row width (*a* and *b*, respectively, in the Figure) were tested 3 ft. (91 cm.) and 1 ft. (30 cm.), 3 ft. and 2 ft. (61 cm.), 4 ft. (122 cm.) and 1 ft., 4 ft. and 2 ft., and 5 ft. (152 cm.) and 1 ft. At each of these combinations the within-row spacing of canes (*c* in the Figure) used were 6, 12, 18, 24, and 30 in. (15, 30, 45, 61 and 76 cm. respectively). Each of these cane arrangements was studied with either one or two plants growing up each cane. This gave a range of plant populations of from 0.13 to 2.0 plants per square foot (1·4 to 21 per square metre) of field and picking was done every three to four days from early August to mid-October.

The results were startling. As population increased from 0·13 to 2·0 plants per square foot yield increased from 4 to over 9 lb. per square yard (2·2 to 5 kilos per square metre). It was also clear that the more even the plant arrangement the higher was the yield. For example, averaged over the within-row spacing 3 ft. (91 cm.) paths with a 2 ft. (61 cm.) twin-row gave a 12 per cent higher yield than a 4 ft. (122 cm.) path and a 1 ft. (30 cm.) twin-row. Further, two plants at each cane gave lower yields than the same population spaced as single plants at each cane, the differences ranging from about 10 per cent at high populations to 20 per cent at lower populations.

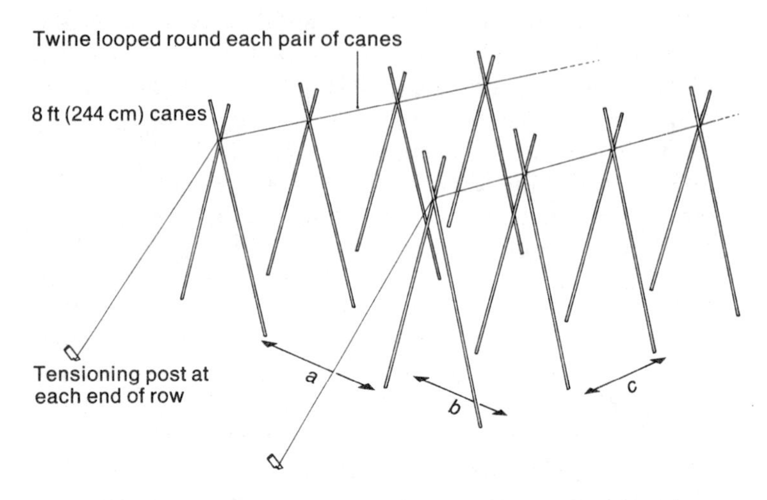

Fig. 1.6. A diagram of the cane system used to support the runner beans grown in the experiments described. (a=path width, b=twin-row spacing, c=within-row spacing.)

High populations delayed the first pick by about three days and the date by which half the eventual crop was picked by up to six days. These effects are very small as compared with the large effects on total yield. No effect on quality was observed.

Commercially an overriding factor governing the conclusions from this work was the cost of the canes and their erection. When this was taken into account the spacing giving maximum profit per unit area was a 3 ft. (91 cm.) path, with twin-rows 2 ft. (61 cm.) apart and canes 12 in. (30 cm.) apart within the row with 2 plants at each cane, giving a population of 1·0 plant per square foot (10·7 per square metre).

In the garden it is unusual to grow more than one twin-row so the path width is somewhat irrelevant. If a system of using 4 ft. (122 cm.) beds with 1 ft. (30 cm.) paths is adopted for the vegetable garden it seems evident that one bed would be devoted to beans using a twin-row spacing of 2 ft. (61 cm.) with each row set in 1 ft. (30 cm.) from the edge of the bed. As canes are expensive some economy can be achieved by using a mixture of canes and strong string. It is more convenient in the garden to use canes to provide a rigid cross member rather than the tensioned wire shown in Fig. 1.6. If 8 ft. (244 cm.) canes are used for this purpose and the row is longer than 8 ft. the canes should be overlapped by at least 18 in. (45 cm.) and lashed together. Each pair of canes must be lashed to the cross member. The within-row canes, for economy, can be spaced at 2 ft. (61 cm.) intervals with string horizontally from the base to base of each cane along each row and similarly from the tip to tip of each cane. Vertical strings can then be tied at 6 in. (15 cm.) spacings along each row running from the top to bottom strings along the same lines as the canes. One plant at each string and cane will give the required population and spacing. It is important to keep on top with the picking as this encourages the production of more pods and so increases total yield. It also ensures you get fewer old and stringy pods.

Broad beans

Research on the spacing of this crop has established that 2 plants per square foot (22 per square metre) will give maximum yields when a single destructive harvest is used. With tall varieties there seemed to be no advantage in using rows closer than 18 in. (46 cm.), but with the shorter, compact varieties, rows 9 in. (23 cm.) apart gave higher yields. When the crop is to be picked on more than one occasion, as is usually the case in the garden, it would be expected that lower populations would be adequate. Thus, $4\frac{1}{2}$ in. (11 cm.) apart in 18 in. (46 cm.) rows would be likely to give maximum yields of tall varieties with 9×9 in. (23×23 cm.) being more suitable for compact varieties.

Peas

As with French beans all the spacing research with this crop has been aimed at producing high yields at one destructive harvest. The varieties used are 'determinate' in that each plant tends to stop growth once it has formed pods at the first two or three flowering positions on the stem and this is unlike the majority of garden varieties. Experiments have shown that 6–8 plants per square foot (64–86 per square metre) in rows not more than 7 in. (18 cm.) apart are likely to give the maximum yield. Results have been much more variable and unpredictable than with other crops in that occasionally populations as low as 4 per square foot (43 per square metre) can yield just as well or even slightly better than higher ones. Thus, we can only talk of 6–8 plants per square foot being more often right than wrong.

The population and row spacing used commercially can be adopted in the garden, leaving the peas to support each other, and harvesting the crop as it is cleared on one occasion only. A succession would then have to be achieved by successional sowings and, of course, only the shorter-stemmed varieties should be used.

The evidence from experiments is that peas respond to being regularly and evenly distributed and that a succession of pods, as required when the crop is to be picked more than once, is encouraged by lowering the population. Thus, if you want to pick-over your crops it would seem reasonable to aim at 4 plants per square foot (43 per square metre) and obtain this by sowing in bands of 3 rows each 4½ in. (11 cm.) apart with the plants 4½ in. apart in each row and with the centre rows 2 ft. (61 cm.) apart.

If the rows run across 4 ft. (122 cm.) beds divided by paths you will be able to pick the crop more easily than if the rows run along such a bed.

INTERCROPPING

With many of the more widely-spaced, slower-growing crops it is possible to use the temporarily-unneeded space between young plants to grow other quick-maturing crops. This can, with advantage, start ahead of the transplanting of such crops as Brussels sprouts. Thus a catch crop of hearted lettuce can be planted in April, leaving spaces for the sprouts to be planted in mid-May. The lettuce will be cleared by mid-June.

Of course, radishes are the ideal quick-growing crop for exploiting the odd no-man's-land, but if all such opportunities were diligently used to grow this crop the average vegetable garden would produce more radishes than it would be prudent to eat.

The golden rule is to avoid the intercrop interfering with the maincrop. If you imagine the intercrop as weeds competing with the maincrop you will appreciate that if too much enthusiasm is applied to intercropping serious damage can be done to the yield of the maincrop. The evidence is that the maincrop would not fully recover from any competitive check caused by early intercropping.

The safe rule is to attempt to exploit only the very centre part of the wide spaces that separate some crop plants early in their growth. Even then such intercropping should be brief and extra fertilizer and water may have to be given to make up for what the intercrop removes.

Of course, using the right spacings for the maincrops implies that you will be on a razor's edge as far as any additional interplant competition is concerned. Thus, not only must you be very careful about intercropping but you must also be alert to prevent nature from intercropping for you with weeds. Getting the better of the weeds in the vegetable garden is one of the keys to success. Remember, it is easier to kill weeds by hoeing when they are very small and that preventing the seeding of weeds is vital. One year's seedling is seven year's weeding!

Table 1.2
A summary guide to spacing your vegetable crops

Crop	Row spacing × spacing within row (in.) (1 in.=2·5 cm.)	Additional comments
Beetroot		
early crops	7×4	For early maturity of 'Detroit' type
maincrop	12×1	For maximum yield of medium-sized roots
Beans		
broad	18×4·5	For tall varieties
	9×9	For compact varieties
dwarf	18×2	For maximum yield as even a pattern of arrangement as possible
runner	24×6	For cane arrangements see p. 26
Brussels sprouts		
regular picking	36×36	To give successional picking as sprouts mature
for freezing	20×20	For single harvest of small sprouts
Cabbage		
spring types	12×4	Remove plants for use to give 12×12 for late-hearted greens
summer	14×14	For small heads
	18×18	For larger heads
winter	18×18	Similar to summer types
Calabrese	12×6	At this spacing both terminal and side spears produced
Carrots		
early crops	6×4	To limit competition and get rapid growth
maincrops	6×1·5	For medium-sized carrots

Crop	Row spacing × spacing within row (in.) (1 in. = 2·5 cm.)	Additional comments
Cauliflower		
early summer	21 × 21	Those varieties maturing June–July
autumn	27 × 27	Those varieties maturing August–November
winter	30 × 30	(Broccoli) Those maturing December–May
mini	9 × 4	To produce curds $1\frac{1}{2}$–$3\frac{1}{2}$ in. in diameter
Celery		
self blanching	11 × 11	To give high yields and good blanching of sticks
	6 × 6	To produce slender hearts
Leeks	12 × 6	Will give maximum yield of normal-sized leeks
Lettuce		
early crops	9 × 8	Triangular patterns for frames or cloches (p. 11)
normal	12 × 12	For butterhead, cabbage, or cos types
leaf	5 × 1	For non-hearted leaf production (p. 11)
Onions		
bulb		
from seed	12 × 1·5	To give medium-sized bulbs
from sets	10 × 2	Will give high yields of medium-sized bulbs
for pickling	12 × 0·5	To give high yields of small bulbs
salad	4 × 1	Can also be sown in a band
Parsnips	12 × 6	For high yields of larger rooted varieties, e.g. 'Offenham'
	7·5 × 3	For smaller roots

Table 1.2 (*cont.*)

Crop	Row spacing × spacing within row (in.) (1 in. = 2·5 cm.)	Additional comments
Peas		
for picking over	4·5 × 4·5 × 18	Sow in three-row bands each row 4·5 in. apart with 18 in. between each band
Potatoes		See Table 2.1 (p. 17)
Shallots	8 × 6	If using small sets
	12 × 6	If using large sets
Tomatoes		
bush types	19 × 19	This close spacing gives more early yield

2 Sowing and planting

More seeds are wasted and more plants are condemned to produce less than their full potential by poor sowing and planting methods than almost any other cultural practice within the gardener's control. In fact, producing the right conditions to encourage rapid germination and growth of seedlings and transplants is the foundation on which high yields of good-quality vegetables are produced. Although this is easier said than done there are certain principles which can be used as a guide by the gardener. If these are appreciated and if the novel methods of sowing described later are followed, then more predictable plant populations with more efficient use of seeds and plants will be obtained, not to mention the increased satisfaction and rewards for your efforts.

GETTING THE RIGHT NUMBER OF PLANTS

Gardeners intuitively appreciate that sowing too many or too few seeds may lead to a crop failure. In fact, getting the right number of plants to grow per square yard for each different vegetable is probably the single most important gardening operation. From seeds this can be very difficult to achieve. For one thing we need to know how many of the seeds will emerge as seedlings. This cannot be done simply by counting the number of seeds we sow because an unknown proportion of them, often as many as half, fail to produce seedlings.

GETTING THE PLANTS TO GROW AT THE RIGHT TIME

The *number* of plants established isn't the only aspect of seedling establishment which is important in ensuring good reliable yields of vegetables. It is also essential to be able to get crops growing at the 'right' time, especially when successional sowings of peas or lettuce are made to provide a continuous supply for the kitchen. A delay in emergence may mean a gap in supply. It is well known that seedlings from one sowing do not all emerge at the same time. Does it matter, then, if it takes a long or short time for all the seedlings to emerge? This will depend on the particular vegetable. For example, with lettuce and cabbage which are spaced widely apart the differences in times of emergence of the individual seedlings, which can be up to ten days, helps to give a continuous supply of heads. This is because the earliest emerging seedlings give the earliest maturing plants and the latest to emerge the latest maturing heads, usually a great advantage to the gardener. However, with other crops which are more closely spaced, such as carrots, onions, or parsnips, a long period of emergence of seedlings can be a disadvantage. This is because the earliest seedlings to emerge, which become the largest, will 'crowd out' the later emerging ones, which then may only produce small worthless roots or bulbs.

TO SOW OR PLANT?

It is usually easier to get a high proportion of the seeds emerging rapidly and all at the same time by raising seedlings under glass in seed trays or soil blocks and then, at a later stage of growth, transplanting them into the garden. This has led to the belief among some gardeners that crops can be established more easily in this way than those sown directly into the garden soil and that the resulting crops will be better. This is not invariably so. Some crops do not readily recover from the damage caused by transplanting. For example, research has shown, even with cauliflowers which are relatively easy to establish, that transplanting may give poorer crops than those raised from seeds sown directly into the garden. Furthermore, some crops such as celery, which are traditionally transplanted, are difficult to establish well either from plants or from directly sowing the seeds.

The reasons why some crops are sown directly whilst others

are best transplanted will be investigated a little later. First, it is useful to understand how to influence the number of seedlings that emerge and establish themselves as plants.

WHY DON'T ALL SEEDS PRODUCE PLANTS?

The successful establishment of healthy seedlings is the end point of a complex process which can be regarded as three separate events. Firstly, the production of 'live' seeds. Secondly, the germination of the seeds which involves the uptake of water to start the growth of the root through the seed coat. Finally, the growth of the root downwards and the growth of the shoot upwards to the soil surface. Seedlings will fail to become established if any one of these events is not completed.

HOW MANY SEEDS ARE 'LIVE'?

One of the reasons why some seeds fail to germinate is because they do not contain a live embryo. This occurs because fertilization of the flower was incomplete or because the fertilized seed did not develop properly. Most of these 'dead' seeds are removed during the preparation of the seeds for sale.

A much more common reason for seeds failing to germinate is because some seeds have abnormal or under-developed embryos. This is common in carrots, parsley, celery, and parsnips, and the seeds need to undergo further development before germination can occur. In these circumstances germination is often delayed and consequently the seed in the soil becomes more vulnerable to disease and pest attack. This is one of the reasons why it is difficult to obtain good emergence from these particular crops and why the seedlings take such a long time to appear. Of course, these particular vegetables are also often sown early in the spring when soil temperatures are low, and this will further delay emergence of the seedlings.

Some seeds fail to germinate, or if they do germinate fail to grow vigorously, because they are diseased. Such seed and seedling diseases are described in Chapter 6.

Some seeds are particularly 'difficult' because they require special conditions before they will germinate. For example, seeds of many celery varieties require light for germination and will not germinate in the dark or if *buried* in the soil. Freshly harvested lettuce seeds will not germinate readily, but good germination can be obtained after a few months of storage. The 'corky' layer around beetroot seeds (which are really fruits) contains compounds which can inhibit germination. This is often rubbed off by the seedsmen to produce a smooth, round 'fruit'. If large amounts of the cork are present they can be soaked in running water for $\frac{1}{2}$–1 hour at a room temperature of about 70°F (21°C) to wash out the compounds which delay germination.

Another common cause of poor germination is because the seeds are too 'old'. Generally, seeds kept for some time germinate less well than freshly harvested seeds; even if some seeds do germinate the seedlings often fail to grow properly because they do not produce a normal root or shoot. Such a statement should immediately raise questions in the gardener's mind: Is germination reduced by even short periods of storage of the seeds, for example, between buying and sowing them? Is it safe to use seeds which have been kept from previous years?

CAN SEEDS BE USED THAT HAVE BEEN STORED FOR ANY LENGTH OF TIME?

It is not sufficiently appreciated that as soon as seeds are harvested from the seeding plant they start to deteriorate and will eventually die if stored for long enough. However, research has shown that there are differences in how quickly different vegetable seeds deteriorate. For example, onion seeds will lose the ability to germinate more rapidly than tomato seeds; the storage life of some common vegetables is shown in Table 2.1. Most seeds will germinate after two or three years' storage in unsealed containers, but if the seeds are kept in warm and humid conditions then they will often fail to germinate even after a few weeks. So, even when new seeds are bought each year, they will deteriorate if stored in warm, moist conditions before sowing. Often the same packet of seeds is used for a number of successional sowings of, say, lettuce and between sowings the seed packet is left on a shelf in the garden shed. Research has shown that seeds stored under these conditions deteriorate rapidly and this form of storage is often the cause of poor germination and seedling emergence for sowings made later in the season.

Table 2.1

The storage life of some common vegetable seeds kept in unsealed containers in a normally heated room

Up to 3 years	Up to 6 years	Up to 9 years	Up to 10 years
Lettuce	Carrot	Cabbage	Tomato
Leek	Cauliflower	Swede	Radish
Onion		Turnip	
Parsnip			

The effects of seed deterioration are twofold. The longer seeds are stored the lower the proportion of seeds which are capable of germinating, but long before this occurs the seeds lose their 'vigour'. This means that although the seeds are capable of germinating under *ideal* conditions they do not under poor conditions. For example, in one study 95 per cent of both 'old' and 'new' carrot seeds germinated under ideal conditions of warmth and moisture, but only 35 per cent of the 'old' and 60 per cent of the 'new' seeds germinated under *cold, moist* conditions. Research has shown that seedlings from 'old' seeds do not grow as well as seedlings from 'new' seeds so giving, for example, later maturity of lettuce and lower yields of carrots.

It is important, therefore, to store seeds properly if they are to be saved from year to year.

HOW SHOULD SEEDS BE STORED?

The loss of the ability to germinate and the loss of vigour can be minimized by proper storage. Remember, high temperature and moist conditions cause rapid deterioration. Therefore, if the seeds are kept dry and stored at low temperatures, they remain alive longer.

In general, for each 9°F (5°C) rise in temperature above 32°F (0°C) the storage life of the seed is halved. So if it is normally ten years at 41°F (5°C) it will be five years at 50°F (10°C), and only one year at 70°F (21°C) (in your kitchen). Also, for each 1 per cent increase in seed moisture content above 5 per cent up to 14 per cent the life of the seed is reduced by half. So seed with a moisture content of 8 per cent will store for ten years but will only store for five years at 9 per cent moisture content. These two effects can be added together so that if the seeds stored at 41°F (5°C) and 8 per cent moisture content remain viable for ten years, then at 50°F (10°C) and 9 per cent moisture content they will only remain viable for two and a half years.

What does this mean in practice? When the seed is mature it will take up water from a moist atmosphere or lose it to a dry one. In a maritime climate like Britain's the relative humidity of the air is rarely below 75 per cent and seeds stored in unsealed containers will usually have a moisture content of approximately 10–15 per cent and so will only store satisfactorily for about one year. This is because the seeds are so moist that moulds develop which destroy the tissues of the seed. If seeds are to be stored from one sowing to the next or from one season to the next then they should not be stored in an unsealed packet in the garden shed. This applies equally to seeds in paper packets and to those in tin foil vacuum-packs which have been opened. However, seeds can be stored for at least three to four years in unopened vacuum packs. Store the seeds in a cool place, such as a cellar which is usually at about 50°F (10°C) throughout most of the year, and keep them dry. This can be done simply by storing seeds in a glass jar or other *transparent* air-tight container with a small amount of cobalt-chloride-treated silica gel (a teaspoonful per ounce (28g.) of seeds) placed in a separate open-topped container (see Fig. 2.1). The silica gel (which can be obtained from most chemists) takes up water from the surrounding atmosphere and so dries it. The moist seeds will then lose moisture to the dry air and the seed moisture content can be reduced to about 8 per cent. It will be possible to maintain it at this level provided the silica gel is taken out and dried after it turns from blue to pink. Drying in an oven for two to three hours will drive off the water and when it is dry (indicated by the change from pink to blue) it can be put back in with the seeds. The silica gel will need to be dried frequently to start with, but when the seed moisture content has reached a balance with water in the air in the container less frequent drying will be necessary.

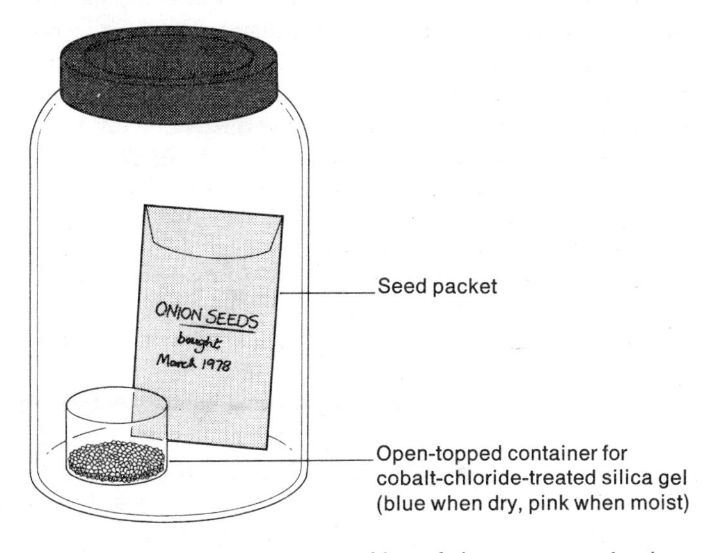

Fig. 2.1. A simple way to store vegetable seeds in a screw-top glass jar.

If seeds are stored in this way they should keep in good condition for three or four years. It will then be possible to buy larger quantities of seeds of, say, a special variety, which is not always readily available, and keep it for several growing seasons.

Although research has shown us how to store seeds to minimize deterioration it is not known with any certainty how to produce seeds of a consistently *high* quality which will germinate well under all conditions; some 'seed' years are good and others are bad. Seedsmen can grade out the smaller seeds, which often germinate less well than the larger ones, but this will not guarantee that all of the seeds in a packet will germinate.

DO SEED PACKETS HAVE TO CONTAIN A CERTAIN PROPORTION OF 'LIVE' SEEDS BY LAW?

It is recognized by all concerned with seed production and marketing that it is difficult to produce consistently good seed and so seeds of nearly all vegetables sold must meet certain *minimum* levels of germination. United Kingdom standards for a number of vegetables are given in Table 2.2. Other countries have similar standards. It is important to realize that these are *minimum* standards and a seed packet will normally contain more 'live' seeds than required by law. The test to determine the

Table 2.2

The U.K. statutory minimum levels of germination for some vegetable seeds

No minimum level required by law	65%	70%	75%	80%
Parsnip	Carrot	Onion	French bean	Broad bean
	Leek	Cauliflower	Lettuce	Runner bean
		Calabrese	Cabbage	Pea
		Radish	Brussels sprouts	Turnip
		Celery	Marrow	Cucumber
		Beet*	Spinach	
			Tomato	

* Except 'Cheltenham Green Top' which is 50 per cent.

number of 'live' seeds is done at the time the seeds are packeted for sale to shops and several months may elapse until the seeds are sown. If the conditions in the shop or home before sowing are poor some of the seeds may die. So, it is best to buy seeds early in the season soon after they are packeted and to store them under good conditions.

The test for determining the number of 'live' seeds is done in a laboratory under *ideal* conditions of light, warmth, and moisture. If these conditions could be reproduced in the garden soil or seedbox all the 'live' seeds should germinate and produce seedlings. However, the soil is hardly ever an ideal medium for germination and seedling emergence and rarely do all the 'live' seeds emerge.

WHY DON'T ALL 'LIVE' SEEDS PRODUCE SEEDLINGS?

There are many reasons why 'live' seeds fail to germinate and produce seedlings in the garden. As well as those already mentioned earlier the following are the most important.

Soil tilth and soil moisture

It is very important to prepare a suitable 'tilth' before sowing so that when the seeds are sown they make good contact with the water held in the pore spaces between the soil particles. The uptake of water into the seed is then rapid and the growth processes begin readily. This uptake of water by the seed is better when a fine rather than a cloddy tilth is prepared. Although a lot of research has been done on the effect of different tilths on water uptake and seedling emergence it is very difficult in practice to produce ideal tilths. The reason is that each soil reacts differently to the various cultivation implements. The same soil will also be affected in different ways, depending on whether it is wet or dry when it is being cultivated. As a general rule, the best tilths for seed sowing are those produced with as little raking of the soil as possible, as this preserves any 'frost mould' on the surface. On loamy, well-structured soils it is relatively easy to produce a good tilth for sowing, but on clays and sandy soils this is more difficult and timely cultivation becomes very important. On clays autumn digging to expose the soil to the natural weathering process will help to produce a good seedbed in the spring. However, some sandy soils dug in the autumn may become compacted by sowing time through the action of rainfall, so making them more difficult to cultivate than those freshly dug. In the long term, tilths on both clays and sands can be greatly improved by incorporating into the soil each year large amounts of compost and other organic matter.

'Slumping' of the soil and surface capping

Some poorly structured sandy soils which contain a high proportion of fine sand, silt, or clay particles can 'slump' under the impact of heavy rain or heavy watering. This is because the fine particles in the soils are washed into the spaces between the larger particles. The soil is then said to have slumped. When it is in this condition and wet, all the air is forced out of the soil and the seeds will be starved of oxygen and will eventually rot. Slumped soils are also very compact, which severely restricts the growth of the shoot and root and this can be recognized by the flattening, thickening, and kinking of the roots and shoots when the seedlings are carefully dug up.

Sometimes this slumping is confined only to the top $\frac{1}{4}$ in. (6 mm.) or so of soil and is referred to as 'capping'. If the soil remains wet under these conditions seedling emergence can occur, but if the soil surface dries this slumped layer forms a hard 'cap' and the seedlings will often fail to emerge. This cap can usually be lifted up from the soil and the seedlings can be found curled round and round underneath (see Fig. 2.2).

The adverse effects of both slumping and capping can be reduced by carrying out the minimum number of cultivations necessary to produce a crumbly tilth and by watering the soil lightly *before* rather than *after* sowing. Some gardeners dribble water into the bottom of the drill before sowing (see Chapter 4). This is a good way of ensuring adequate moisture for germination without damaging the soil structure. If watering is necessary after sowing then small amounts of water should be applied frequently and with a fine rose. If the soil does form a cap it should be kept wet by frequent waterings to reduce its mechanical strength.

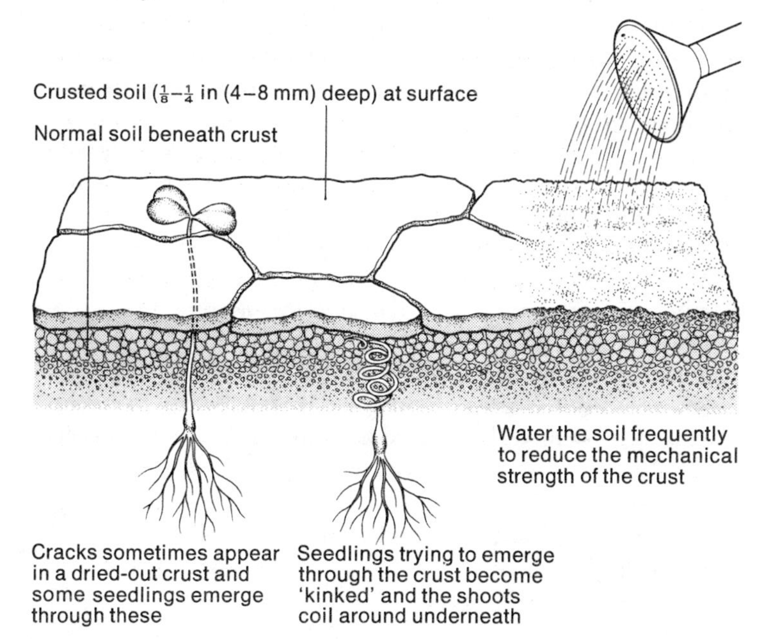

Fig. 2.2. Soil crusting or capping.

If the sown area is small enough or if widely spaced crops are 'spot' sown on soils prone to capping it is worth considering replacing the soil covering with a non-capping material, such as peat, to encourage rapid seedling growth to the surface.

Rolling

It is claimed that treading or rolling the seedbed is beneficial for seedling establishment because it improves the tilth and the contact between the seed and the soil water. This will depend on whether it is done before or after sowing and how moist the soil is at the time of consolidation. In cloddy soils rolling the seedbed before sowing will help to produce a good tilth, but only if the soil is at the right moisture content so that the clods crumble. This can only be judged for your soil by experience. Normally rolling should not be necessary. In wet soil its only action will be to compact the soil, which will hinder the growth of the shoot and root. Rolling seedbeds after sowing, particularly if the surface of the soil is moist, can predispose the soil to form a cap if the seedbed dries out rapidly afterwards. Generally, treading or rolling the seedbed is not recommended.

Sowing depth

To ensure good contact with the soil seeds need to be sown in a drill and covered with soil. However, this can bring other problems because the seedling has to live on its own food reserves until it emerges from the soil and begins to manufacture its own food. If the seed is sown too deeply its reserves

Table 2.3
A guide to the depth of sowing for some common vegetables

1½–2 in. (38–50 mm.)	1–1½ in. (25–38 mm.)	¾–1 in. (19–25 mm.)	½–¾ in. (12–19 mm.)
Broad beans	Peas	Broccoli	Beetroot
French beans	Sweet corn	Brussels sprouts	Carrots
Runner beans		Cabbages	Leeks
		Cauliflowers	Lettuce
		Cucumbers	Onions
		Marrows	Parsley
		Radishes	Parsnips
		Swedes	Spinach
		Tomatoes	
		Turnips	

may not be sufficient to sustain growth to the soil surface.

Small seeds have small reserves and therefore must not be sown too deeply. For example, the optimum depth of sowing lettuce in moist soil is about $\frac{1}{2}$–$\frac{3}{4}$ in. (12–19 mm.), whereas for beans, which are 500–1000 times heavier, it is about $1\frac{1}{2}$–2 in. (38–50 mm.). Suitable sowing depths for other vegetable seeds are shown in Table 2.3.

In practice, it is not easy to draw drills in the soil at the same depth along their length using a rake or hoe, and some seeds will be sown more or less deeply than others. Although the ones sown too deeply germinate they will not emerge and an uneven depth of sowing is a fairly common cause for only a small percentage of the seedlings emerging. Better control of the depth of sowing can be obtained by using a planting board and a simple model is shown in Fig. 2.3. Using this board and a V-shaped scraper, drills can be made at the correct depth along the whole length of row. The board can also be marked off to aid the precise spacing of seeds along the row.

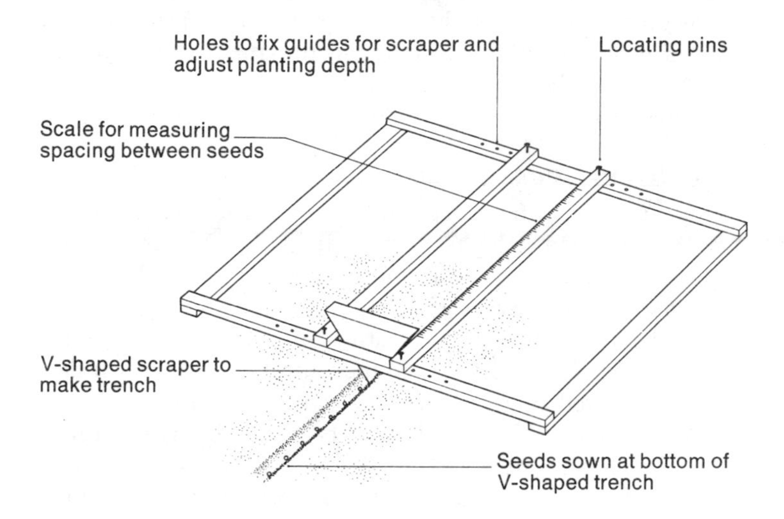

Fig. 2.3. Sowing frame to control the depth of sowing.

Variation in the depth of sowing is also a major cause of differences in the time of seedlings emerging, because those sown nearer the surface will usually emerge first providing, of course, that the soil is not too dry. Another cause of variation in time of emergence is if the seeds start to absorb water, and therefore start to germinate, at different times. This can occur, for example, if some seeds are sown in a seedbed with a poor, cloddy tilth.

Fertilizer practice

Inorganic fertilizers are usually applied to the seedbed before sowing and forked or raked in, but if large amounts of nitrate are applied at this time seedling emergence will be reduced, particularly under dry conditions. This is because the fertilizer prevents germination and reduces growth of the small seedling root. This effect is particularly marked in small-seeded crops such as lettuce and carrots, but it also affects larger-seeded ones such as beans. Provided the soil is kept moist by watering each day the damaging effects of excess fertilizer can be minimized. However, it is probably better to apply a small proportion of the nitrogenous fertilizer required by the crop (up to one-third) to the seedbed before sowing and the rest later when emergence is complete.

Soil temperatures

In general, seeds take longer to germinate and germinate less well at low than at higher temperatures. For example, emergence may take up to four or more weeks for peas and parsnips when sown in the early spring but only a week or ten days when sown later after the soil has warmed up. There are wide differences between vegetables in the temperatures at which they germinate well and it is useful to know some of these.

Vegetables such as cauliflowers, cabbages, Brussels sprouts, turnips, radishes, peas, and broad beans will germinate successfully and rapidly from temperatures as low as 41°F (5°C) to as high as 90°F (32°C) and so there are few restrictions on the time of year when seeds can be sown to establish plants. Other vegetables are not so tolerant to very low temperatures or to high temperatures. For example, leeks and onions will not germinate well above 70–75°F (21–24°C), nor do they germinate well at temperatures below 45°F (7°C). So from early sowings seedlings usually take a long time to emerge. It may also be difficult to establish the new Japanese types of overwintering onions which are sown in August. Often, at this time of the year soil temperatures reach 70–75°F (21–24°C), which may reduce seedling emergence. If soil temperatures at seed depth are as high as this it could be prudent to lower the temperature by watering the soil frequently or by covering the soil with a white reflective material such as white polythene or polystyrene.

Another vegetable which will only germinate well within a narrow range of soil temperature is celery, germination being best at temperatures between 50° and 66°F (10° and 19°C). Although lettuce seed germinates well at low temperatures (it will in fact germinate in ice) most varieties of the butterhead (smooth) types will not germinate when soil temperatures are above 77°F (25°C). If the soil cools down after a hot spell germination will occur, but only after a considerable delay and when it does occur the seedlings emerge over a long period of time. As soil temperatures above 77°F occur frequently in late spring and summer, even in England, this is one of the major reasons for poor germination and emergence, resulting in unpredictable supplies of lettuce during the late summer and early autumn. Crisphead varieties of lettuce will germinate well even at 85°F (29°C) and so there would be less risk of poor germination if these were used instead. If a crisp variety is not an acceptable alternative then soil temperatures can be reduced by watering or using a white reflective material to cover the soil or seedbox. Alternatively, sowing at a particular time of day will aid germination when soil temperatures are high. Research has shown that sowing between 2 and 4 p.m. gives better emergence than sowing in the morning or early evening. This is because the temperature-sensitive phase of the germination process is completed at night when soil temperatures are lower.

In contrast, carrots, parsnips, and red-beet germinate well

at high temperatures but not at temperatures below about 45°F (7°C). Germination does occur, but it is very slow and this, in part, accounts for the very long time, often four to six weeks, that it takes for these seeds to emerge from early spring sowings. One of the ways of getting earlier and more rapid emergence is by pre-germinating and fluid sowing the seeds and this will be described later.

Some vegetables, notably sweet corn, tomatoes, cucumbers, courgettes, marrows, peppers, runner beans, and French beans, are particularly sensitive to low soil temperatures. For example, cucumbers and marrows will not germinate satisfactorily below 56°F (13°C) whilst peppers require soil temperatures above 60°F (15°C) for successful germination. This temperature is not reached on most soils in the United Kingdom until early June, which is too late to sow pepper seeds to get a crop. Other cold-temperature-sensitive crops such as French and runner beans, tomatoes, and sweet corn require soil temperatures above 50–54°F (10–12°C) for successful and rapid germination.

It is important to realize with these low-temperature-sensitive crops that if the seeds imbibe water in cold conditions, the growth of the seedlings which eventually emerge is permanently impaired and yields may also be reduced. So the seeds should not be sown until there is little risk of low soil temperatures occurring. This means that for sweet corn and beans sowing should be delayed until soil temperatures reach 50°F (10°C). (In the South of England this occurs in about late April to early May and in the North of England a week or two later.) For cucumbers and marrows, sowing should be delayed for a further two weeks. Slightly earlier sowing can be done if the soil temperature is artificially raised by protecting the seedbed and the growing crop with cloches, low plastic tunnels, or clear plastic mulches (see Fig. 2.4). This not only ensures rapid emergence but will advance the time of maturity.

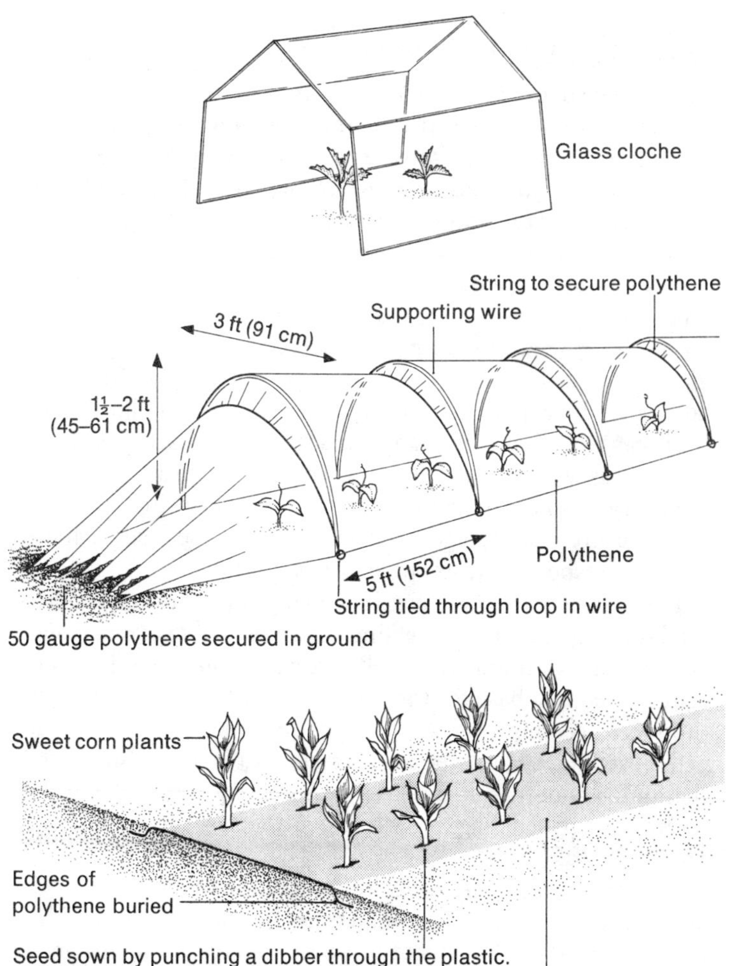

Fig. 2.4. Methods of warming up the soil for cold-sensitive vegetables. Cloches and plastic tunnels are good for French beans, courgettes, cucumbers, and bush tomatoes; clear plastic mulches can also be used for these crops and are also useful for sweet corn.

HOW CAN THE NUMBER OF SEEDLINGS THAT EVENTUALLY EMERGE BE PREDICTED?

Only rarely does the right combination of conditions in the garden seedbed occur so that all the 'live' seeds produce seedlings. These conditions almost always arise by chance and rarely can they be produced to order even though care has been taken with seedbed preparation. Gardeners usually accept philosophically that emergence is unpredictable, nevertheless it is possible to improve on this situation. Sowings can usually be classified into those which give 'poor' or 'good' emergence. If the gardener knew, for example, whether 25, 50, or 75 per cent of the 'live' seeds sown produced plants under particular conditions then, by experience, he could begin to predict how many seedlings would emerge at any one sowing. For example, if 100 'live' seeds were sown and on average only 50 came up for February sowings compared with 75 for April sowings he could sow in future years $100/50 \times 100$ and $100/75 \times 100$ seeds at those sowings, respectively, to obtain 100 seedlings.

It is instructive to see just how many seeds are required to establish, say, 40 lettuce seedlings in a row 5 ft. (152 cm.) long. If the statutory minimum level of germination which is 75 per cent is used as a base for the calculation, and all the live seeds produced seedlings, we would need $(40 \times 100)/75 = 53$ seeds to get 40 seedlings. However, in experiments it is frequently found that only about half of the 'live' seeds emerge and so in order to produce 40 seedlings we would need to sow $(40 \times 100)/75 \times 100/50$ or 106 seeds.

In summary, sowing seeds into soil does not always provide an ideal environment for the germination and growth of the seedlings, but establishment can be improved by attention to the following points:

Prepare a good tilth before sowing.

Apply to the seedbed only one-third of the nitrogen fertilizer required to grow the crop, applying the rest later.

Sow the seeds at the right depth for each crop using a sowing board (Fig. 2.3).

Water the bottom of the drill before sowing; if the soil becomes dry after sowing, water frequently with small amounts using a fine rose.

Delay the sowing of temperature-sensitive seeds until the soil has warmed up or warm the soil by protecting it with cloches or clear polythene.

Prevent the formation of a soil crust or cap by replacing the soil covering with a compost mixture.

FLUID SOWING—A NOVEL SOWING METHOD

Even when considerable care is taken to prepare a seedbed it is still difficult to provide the ideal conditions of moisture, soil tilth, and temperature by traditional garden cultivations alone in order to get the right number of seedlings growing at the right time. So techniques of seed sowing have been developed which eliminate the effects of weather and soil conditions on seed germination and minimize their effects on subsequent growth of the shoot to the soil surface. It is especially useful for getting rapid and uniform emergence of seedlings early in the season when the soil is cold, but it also gives advantages in other circumstances.

This technique is called fluid sowing. The seeds are germinated indoors under *ideal conditions* and the germinated seeds are then sown into the soil in a jelly to protect them from damage. A simple example of how the effects of the soil environment can be eliminated is where soil temperatures greater than 77°F (25°C) prevent germination of lettuce seeds sown into the soil during hot sunny weather. However, provided the seeds are *germinated* under ideal conditions (in temperatures below 77°F) *before sowing* and then fluid sown, seedlings will come up readily even with soil temperatures above 77°F because such high temperatures do not prevent the subsequent growth of the young shoot.

Fluid sowing of pre-germinated seeds has now been tested experimentally on a wide range of vegetables and for all of them

Table 2.4

A summary of likely benefits from fluid-sowing pregerminated seeds compared with sowing dry seeds

	Number of days earlier emergence for fluid-sown pregerminated seeds compared with dry seeds for		Other benefits
	early season sowings	late season sowings	
Carrot	12	5	Gives higher yields of early carrots
Celery	21	1	Better emergence in seedboxes
Lettuce	8	1	Better emergence during mid-summer
Onion (bulb)	15	4	Higher yields
Onion (salad)	15	7	Ready for pulling earlier
Parsley	21	14	Better and earlier emergence
Parsnip	18	4	Better emergence and ready earlier
Tomato	20	7	Mature crops can be produced directly from seeds sown in the garden soil in countries where the season is normally too short

it gives earlier emergence, particularly with those crops such as carrots and parsnips which are normally slow to germinate, or those such as tomatoes which germinate only slowly at low temperatures. In fact, with tomatoes emergence and growth are advanced so much that it is possible to produce ripe crops by sowing directly into the garden soil even where the season has been regarded as being too short to get a satisfactory crop, for example in the United Kingdom.

This method will also give higher and more predictable emergence of seedlings than dry seeds and particularly with 'difficult' seeds like celery. In lettuce it will also give more uniform emergence, all the seedlings coming up at approximately the same time. The advantages found from using this technique with other vegetable crops are listed in Table 2.4. The technique is now starting to be used on a commercial scale in several countries.

The method can be easily adapted for the home gardener, using kitchen containers and utensils. The following steps should be followed:

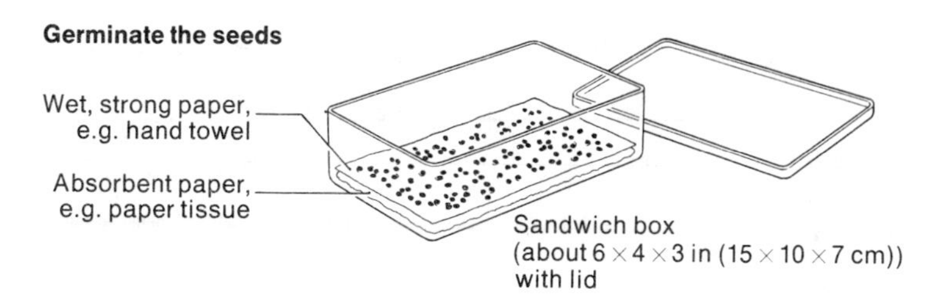

1. Line the bottom of a sandwich box, empty seed tray, or other container (a margarine container would do) with a $\frac{1}{10}$ in. (2–3 mm.) thick layer of absorbent, unmedicated paper such as paper tissues; cover this layer with a 'wet-strong' paper such as a hand towel.
2. Sprinkle water on to the paper until it is thoroughly wet and then pour away the excess.

3. Sprinkle the seeds evenly over the paper without completely covering the surface and avoid clumping of the seeds. It is important with celery and parsnips to spread these very thinly—about six seeds to every square inch (one seed per square centimetre).
4. Put the lid on the container and keep the seeds at a temperature of about 70°F (21°C) for all the common vegetables. Remember that celery must be kept in the *light* otherwise germination will be poor. Some seeds like lettuce germinate in less than twenty-four hours at 70°F so it is important to look at the seeds frequently to see if they have germinated. If fluid-sowing of pre-germinated seeds is to be successful the roots should not be allowed to grow too long—up to $\frac{1}{5}$ in.

Table 2.5

The approximate time required at normal room temperature for the pre-germination of seeds for fluid sowing and the approximate percentage germinated at this time

	Time required (days)	Germination (%)
Bean (French and runner)	3	80
Beetroot	6	70
Carrot	4	50
Celery	10	50
Cucumber/Marrow	1½	70
Leek	5	50
Lettuce	1	80
Onion	5	50
Parsley	7	30
Parsnip	7	30–40
Sweet corn	1½	80
Tomato	5	80

(5 mm.) for most vegetables, but lettuce should be no longer than $\frac{1}{10}$ in. (2 mm.). The approximate time needed to produce a root of the right size at the time when the germinated seeds should be sown is shown in Table 2.5 for some common vegetables. These figures should only be used as a guide to the time required because not every lot of seed behaves in the same way.

Germinating the seeds in this way provides a unique opportunity to see before sowing how many, or how few, of the seeds germinate. If the number of germinated seeds is very much lower than the figures given in Table 2.5 after they have been kept in the box for the required length of time they are probably not worth sowing. However, before a final decision is made whether to discard the seeds or not, wait two or three days longer to see if any more germinate.

5. When most of the seeds are at the right stage they may be sown, but if this is not a convenient time, perhaps because the soil is too wet after rain, the seeds in the container can be placed for a day or two in a domestic refrigerator *but the seeds must not be frozen*. Tomatoes, sweet corn, and beans are an exception to this suggestion and should not be stored at temperatures below 42°F (6°C). The seeds can be removed from the box by washing them into a fine-mesh (preferably plastic mesh) strainer.

Wash the germinated seeds off the paper

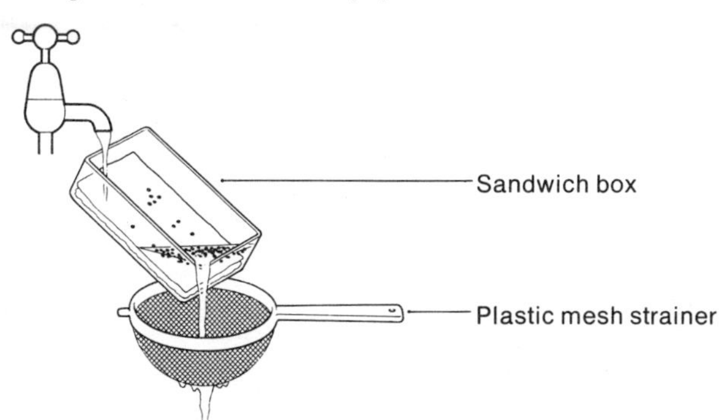

6. The carrier jelly can be made from several compounds, including cellulose-based wallpaper pastes *which do not contain fungicides*. For example, wallpaper paste can be mixed at half normal strength to make a suitable jelly. When the jelly has thickened sprinkle the seeds on to half the required quantity for sowing and stir in using your fingers. Then add the remainder. The seeds should be well mixed in the jelly and should not sink when the mixture is left to stand. If they do the mixture is not thick enough.
7. Fill a wide-nozzled (at least $\frac{1}{4}$ in. (6 mm.) diameter) cake-icing syringe with the mixture and extrude it into moist soil in the seed drill or into a seedbox containing compost. The mixture should be extruded at a rate of 30 teaspoonfuls to 10 ft. (2·5 metres) of row so this amount of jelly should contain about the right number of germinated seeds. Instead of using a syringe, sowing can be done with a polythene bag. After filling the bag and holding the top closed, cut off a small piece of one corner and extrude the mixture through the hole by squeezing as shown in the figure.

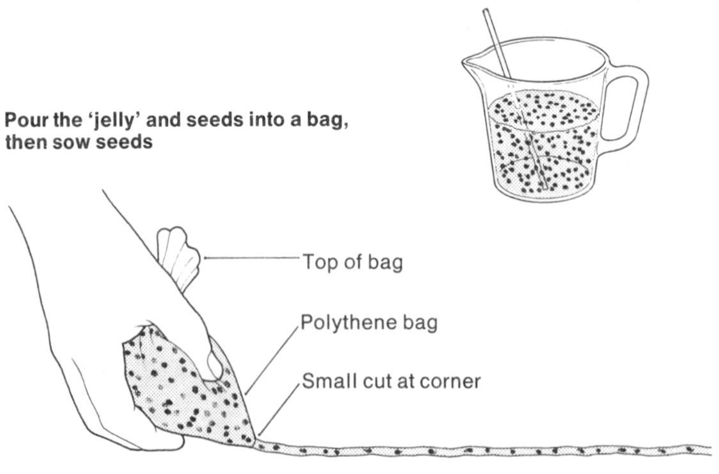

8. After sowing, the jelly containing the seeds needs to be covered in the normal way to prevent it from drying out because the jelly may form a hard film and trap the seeds. Sowing into very dry soil will produce a similar effect and so dry soil should be watered in the seed drill before sowing. For some widely spaced crops it is possible to replace the soil covering by a water-holding material such as vermiculite or a peat compost which does not restrict the growth of shoot to the same extent as soil. Research has shown that higher, more rapid, and predictable emergence can be obtained with these covering materials, but they must be kept moist in drying weather.

DO 'PELLETED' SEEDS GIVE BETTER EMERGENCE THAN NATURAL SEEDS?

It is now possible for gardeners to buy pelleted seeds of many vegetables. One of the advantages of using pelleted seeds is that because they are larger than the natural, uncoated seeds accurate placement of the seeds in the soil is easier. However, it should be remembered that 'pelleting' of seeds was originally carried out to make angular-shaped seeds round so that they could be sown at precise spacings with precision drills. There is no experimental evidence which conclusively shows that seedling emergence is any better from pelleted than from natural seeds. Indeed it can often be far worse.

WHY ARE SOME CROPS SOWN AND OTHERS TRANSPLANTED?

Some vegetable crops are by tradition sown, others transplanted, but traditions vary in different parts of the world. For example, in Japan nearly all the vegetables are transplanted even for commercial production while in England vegetables are both *sown* and *transplanted*.

One of the advantages of transplanting is that it shortens the growing period in the garden so allowing more crops to be grown on the same land than could be achieved by sowing directly into the soil. For example, the same piece of land could support three or four successive crops of transplanted lettuce in a season but only one or two crops from directly sown crops.

Transplanting can also give higher yields, particularly of heat-loving vegetables such as outdoor tomatoes. Indeed, with tomatoes it is the only sure way in a cool climate of obtaining good *early* yields outdoors. Bulb onions from sown crops do not always mature early enough to be dried easily for storage over the winter and so crops are often raised from sets or plants. More importantly, sets and plants give consistently higher yields than from crops sown directly in the soil in early spring.

Transplanting is also used for crops whose seeds are difficult to germinate in the garden. Celery is a good example because the conditions necessary for good germination are more easily obtained in seedboxes than in the garden soil. It is also used for crops whose seeds are expensive such as the newer F_1-hybrid sprouts.

Transplanting is more common for those plants which are widely spaced like cabbage and other brassicas because a small plant-raising area will provide plants for a large area of ground, whereas for close-spaced crops, like onions, a very large plant-raising area would be needed relative to the space for growing them to maturity. In addition, it would be tedious to plant out the very large numbers of plants required.

The major disadvantage of transplanting is that the plant is damaged, particularly the roots, in moving it from the seed tray or seedbed to the soil. Because the shoot continues to lose water by transpiration and because the damaged root system is unable to absorb sufficient water from the soil to replenish this loss the plant wilts. When this happens the plant is unable to produce its own food material and growth is temporarily, but severely, reduced. Rapid recovery from this transplanting shock is essential if yields are not to be reduced and maturity delayed. Some vegetables such as cabbage, cauliflower, Brussels sprouts, and tomatoes recover from transplanting more readily than, say, ridge cucumbers or sweet corn and the yielding potential is less affected. The former vegetables can be readily established from plants pulled from a seedbed (so called 'bare-root' transplants which have little or no soil retained around the roots), but the latter vegetables need to be grown in pots or compressed soil or compost 'blocks' and handled carefully in the field. Research has shown that those plants which can prevent excessive loss of water from their tissues and produce new roots readily recover most rapidly.

WHAT PRACTICAL STEPS CAN BE TAKEN TO PREVENT THE TRANSPLANTING CHECK?

When plants are transplanted from seedboxes or a seedbed at least 25–50 per cent of the root system is lost or damaged. Experiments have shown that removing this amount of roots substantially delays recovery from the transplanting shock. So it is important to reduce root damage to a minimum and promote the growth of new roots as quickly as possible.

The amount of damage will obviously be less for plants raised in soil or peat blocks or peat pots than for bare root plants because the intact block of soil can be planted with the minimum of root disturbance. So raising plants in pots or soil blocks is a good way of reducing the transplanting check. Bare root transplants will suffer considerable damage when the plants are pulled for transplanting unless care is taken in the preparation of the seedbed where the plants will be raised. The soil should be well prepared, if possible digging in peat or compost to produce a good friable soil to encourage the growth of a good root system. Sowing too thickly will give too close a spacing and will increase the intertwining of roots which will be damaged when the plants are separated. In the garden 10 plants per square foot (100 plants per square metre) will be a useful population and will reduce root intertwining. Less damage will be done to the plants if they are loosened gently with a fork as they are being lifted. This will also preserve the soil around the roots, so preventing them drying out rapidly. Watering the ground prior to pulling the plants will also help to reduce damage.

During the transplanting of forest trees the roots are often dipped in a water-based gel usually made from sodium alginate to reduce water loss from the roots and to prevent them drying out. This improves re-establishment in trees, but no research work has been carried out on vegetables to demonstrate conclusively its effectiveness.

It is claimed that the degree of transplanting check can be reduced if the plants are 'hardened' prior to lifting. This involves reducing the amount of water to the plants, reducing the growing temperatures, or both during the last week or so of growth before planting out. Hardened tomato plants develop new roots faster probably because 'hardening' increases the sugar content of the leaves and the amount of wax on the leaves. This would aid renewal of root growth after planting and prevent water loss through the leaf surface, but experiments show that 'hardening' does not consistently improve transplant establishment for other vegetables.

Nearly all vegetables can be transplanted at the early seedling stage with little or no check to growth, but the older the plant the greater is the transplanting check. The check is particularly marked in vegetables which are flowering. This is because the plant does not produce new roots as readily at this stage of growth (see p. 97) as younger plants and so recovery from shock and the effects on yield are greater. For example, tomato plants transplanted at the flowering stage give, on average, 20 per cent lower yields than younger plants. Cauliflowers produced from transplants eight to twelve weeks old are delayed in maturity and give smaller and lower quality curds compared with those produced from five- to six-week-old transplants. The ideal growing conditions and length of the growing period for a number of vegetables are shown in Table 2.6.

It is traditional practice to clip off the leaves or parts of the leaves of transplants at, or just after, transplanting. It is generally believed that this prevents water loss, but there is no conclusive experimental evidence to support this practice and it could be detrimental because clipping off part of the leaves will reduce the ability of the plant to make enough food reserves for a new root system. It is common commercial practice in certain parts of the world to clip the leaves off the plants in the seedbeds *prior* to transplanting. This is done to promote a uniform plant size by 'holding back' the larger plants. It is also claimed with celery and tomatoes that this aids recovery of the transplant if the clipping is carried out two to four days before transplanting takes place, but not all the experimental evidence supports this claim.

Water loss can also be reduced by planting out on dull rather than sunny days and also planting out late in the day. Even when this is done it will still be necessary to water the plants frequently until they are 'standing up' and showing no signs of wilting early in the morning. It is not necessary to water all the soil surface, water can be more efficiently used by watering a small area around each plant (see Chapter 4). Although the maintenance of the water status of the plant after transplanting is most important the damaged root system is also less capable of obtaining nutrients from the soil. So for all vegetable transplants, it is beneficial to provide these in a concentrated and readily available form close to the root system. This can be done by applying a 'starter' fertilizer at the rate of ½ pint (about 600 ml.) to the base of each plant. These are commonly made with high analysis soluble fertilizers. A suitable starter can be made by adding 1¼ oz. (35 g.) and 2 oz. (56 g.) of 'Kaynitro' and triple superphosphate, respectively, to 2½ gallons (11½ litres) of water. This should be thoroughly mixed up and left overnight. Any insoluble material left in the bottom of the container can then be discarded.

In summary, transplanting causes damage to the root system. Rapid recovery from this damage can be promoted by attention to the following:

Prepare the soil carefully and dig in compost or peat where the plants are to be raised.

Allow adequate growing space for 'bare root' plants to reduce root intertwining.

Plant young and not old plants.
Lift the plants carefully from the seedbed.
Plant out on dull days and water the plants frequently until they are established.
Apply a 'starter' fertilizer.
Grow transplants in 'blocks' or peat pots.

Table 2.6

A guide to the growing periods and conditions for transplant raising

	Temp. °F(°C)		Length of growing time (weeks)
	Day	Night	
Cabbages	60–68 (15–20)	50–60 (10–15)	5–7
Cauliflowers	60–68	54–60 (12–15)	5–7
Broccoli	60–68	50–60	5–7
Brussels sprouts	60–68	50–60	5–7
Lettuce	60–68	50–60	6
Onions	60–68	45–54 (7–12)	4–10*
Celery	63–72 (17–22)	54–63 (12–17)	5–10*
Tomatoes	68–77 (20–25)	60–63 (15–17)	6
Cucumbers	68–88 (20–30)	60–68 (15–20)	4–5
Marrows	68–77	60–63 (15–17)	4–5
Courgettes	68–77	60–63	4–5
Peppers	68–77	60–63	9–10

* Shorter period for mid-season crops and longer period for early crops.

3 Plant foods and feeding

Discussions between gardeners on the feeding of plants can often become heated because they hold strong views on the subject. Yet for the majority of gardeners these views are based on accumulated experience over the years under particular conditions, and provided acceptable crops are obtained he regards the results as vindication of his feeding policy. Rarely does he have the time, space, or inclination to compare alternative methods or levels of manuring different crops. Although experience is a sound way of learning, the process can be long and difficult. By understanding the principles of feeding crops and by knowing about the general and relative responsiveness of different crops to fertilizers based on the results of many experiments, this process can be made shorter and easier for the gardener.

PLANT NUTRITION

All plants require a number of 'foods' to be able to grow, some of these being obtained from the atmosphere and others from the soil. Carbon is obtained from carbon dioxide in the air and water is absorbed through the roots. The carbon, hydrogen, and oxygen are combined in a series of complex processes in the plant to produce the carbohydrates which form the necessary basic building materials for plant tissue. The carbohydrates, in turn, are converted into more complex substances such as proteins and fats by combination with mineral nutrients such as nitrogen, phosphorus, and sulphur which are taken up by the roots.

So, as well as ensuring that essential sunlight is not excluded from plants by heavy shading and that there is sufficient moisture available for the plants to grow, the gardener must also ensure that the mineral nutrients necessary for rapid growth are available in the soil. Usually nutrients are added to the soil in the form of fertilizers, composts, or other manures, and in this chapter the nutrients required by various plants are discussed and indications are given of how to decide on the amount of fertilizer or manure to apply to obtain good yields of the different vegetables grown in the garden.

The essential nutrients

Besides carbon, hydrogen, and oxygen which make up about 95 per cent of plant tissue, there are certain nutrient elements which are essential for growth. Some are required in relatively large quantities, such as nitrogen needed for producing proteins and magnesium which is a constituent of the green pigment chlorophyll. Other nutrients are required in much smaller amounts but are nevertheless just as vital for growth.

It is generally considered that there are twelve essential nutrients taken up from the soil. Six are referred to as major elements and six as minor or trace elements. These are listed below:

Major elements		**Minor elements**	
Nitrogen	(N)	Iron	(Fe)
Phosphorus	(P)	Manganese	(Mn)
Potassium	(K)	Boron	(B)
Calcium	(Ca)	Zinc	(Zn)
Magnesium	(Mg)	Copper	(Cu)
Sulphur	(S)	Molybdenum	(Mo)

The elements required for plant growth occur naturally in most soils, being derived from the weathering of minerals or the atmosphere in rainwater (nitrogen), or by the decay of plant material. Soil micro-organisms help to decompose slowly organic matter or humus and produce the simplest forms of nitrogen and other nutrients which can be absorbed by the roots of plants. At the same time, nutrients are lost from the soil by leaching, this being a particularly large source of the loss of nitrogen in coarse or sandy soils.

Crops take up the nutrients from the soil in considerable quantities and the natural cycle of the return of decaying plant waste to the soil is broken if the crop is removed for consumption. It is therefore necessary to supply extra nutrients as manure or fertilizers to make up for these losses and to ensure good growth of succeeding crops.

The soil as a nutrient source

As well as acting as a base for the plant roots to grow, the soil provides the main source of the plant's nutrients, the majority of which are dissolved in the soil water. A good soil structure is, therefore, vitally important for the plant to be able to develop a good root system to obtain the water and nutrients it needs and so to grow satisfactorily. Some methods of obtaining good soil-structure are briefly discussed in Chapter 4 (p. 92).

The roots of some vegetables have been shown to grow far deeper in the soil than many gardeners realize. For example, in well-drained porous soils, roots of some crops such as parsnips and broad beans have penetrated 4 ft. (122 cm.) or more. However, feeding of most crops is done by applying fertilizers or digging composts in the top 12 in. (30 cm.) of soil. This produces more favourable conditions for root growth in this top layer, but deeper-rooting enables the plant to obtain nutrients from the lower layers of soil containing moisture, especially in drought periods. For this reason, it is worthwhile digging in compost or manures as deeply as possible.

SOIL ACIDITY AND THE USE OF LIME

Too often it is evident that some gardeners are under the impression that a good dressing of lime all over their vegetable patch each autumn is an essential for good cropping. At the other extreme, there are those who never consider liming their gardens.

Depending principally on its content of calcium, a soil can be acid, neutral, or alkaline, the former condition indicating that alkaline materials such as calcium salts are in short supply and the latter that abundant alkaline salts are present (for example, in a calcareous or chalky soil). It is well known that some plants prefer acid soils whilst others prefer more alkaline conditions. Two things are therefore important for the gardener to know—firstly, whether his soil is acid or alkaline, and secondly, the preferred level of acidity or alkalinity of the crops he intends to grow.

Soil pH

The acidity–alkalinity of a soil is usually measured on the pH scale. The theory and basis of this scale is beyond the scope of this book, but a value of 7 indicates the neutral mid-point of the scale, which is the pH of chemically pure water. The pH of soils usually ranges from about 3 in acid moorlands to about 8 or 9 in soils containing high proportions of chalk or limestone (calcium carbonate). The majority of soils have a pH in the range 4·5–7·5.

A satisfactory estimate of soil pH can be obtained with an indicator dye which changes colour with the degree of acidity or alkalinity of soil solution. This method forms the basis of several soil test-kits which are on the market. Usually a small amount of soil is shaken with distilled water and the indicator added; after settling, the colour of the clear solution is compared with a standard colour chart.

The acidity or alkalinity of a soil influences the availability of some nutrient elements to plants. A deficiency of calcium can occur at low pH values, but it is difficult to distinguish the symptoms from those of high concentrations of aluminium and manganese which become soluble in acid conditions. Molybdenum deficiency is more prevalent when the soil is acid and is best known for producing 'whiptail' in cauliflower, with a characteristic deformation of the leaves and death of the growing point (see Fig. 3.1). At high pH values, deficiencies of iron, manganese, and boron are common, since these trace elements are rendered insoluble in the soil and therefore become unavailable to plants. The symptoms of these deficiencies are described later in this chapter.

Fig. 3.1. Young cauliflower plant showing symptoms of molybdenum deficiency. (Note that the main growing point has not developed; the young leaves are characteristically stunted.)

It is impossible to give a precise guide as to the soil pH required by each of the different vegetables. Crops grown on organic soils (such as the peats) will grow perfectly well at lower pH values than the same crops grown on mineral soils. There is considerable variation depending on soil type, but it is generally accepted that vegetable crops are best grown on mineral soils with pH values between 6·0 and 6·5. On a peaty soil, plant growth will generally be satisfactory with a pH value of 5·5–6·0.

Plant sensitivity to pH and lime

The various species of plant grown in the garden differ in the range of pH values within which they will grow successfully; an attempt is made to indicate these ranges below:

Soil pH values preferred by vegetable species

pH 5·0–6·0	pH 5·5–7·0		pH 6·5–7·5
Potato	Broad bean	Parsnip	Asparagus
Rhubarb	Brussels sprouts	Peas	Beetroot
	Cabbage	Radish	Carrot
	Cucumber	Swede	Cauliflower
	French bean	Sweet corn	Celery
	Marrow	Tomato	Leek
	Parsley	Turnip	Lettuce
			Onion
			Spinach

How much lime?

Adjustment of the soil pH for individual crops is impossible as it often takes several seasons to correct soil acidity by liming. The best policy is to adjust garden soils to about 6·5 (5·8 if a peat soil) at which all vegetables should grow successfully.

A simple pH test will indicate whether a soil needs liming but the amounts required are dependent on the soil type. A sandy soil with a course texture requires about ½ lb., a loam 1 lb., and a clay or organic soil 1½ lb. per square yard (270, 540, and 810 g. per square metre respectively) of ground limestone to raise the soil with a coarse texture requires about ½ lb., a loam 1 lb., and to use several annual dressings of smaller quantities rather than to use a large amount on one occasion. Hydrated lime can also be used, the effects being similar except that hydrated lime reacts more quickly with the soil but is less pleasant to handle.

When to apply lime

The purpose of liming is to neutralize the acidity in the soil, but this is a process that is not completed quickly. In the case of hydrated lime, damage to crops can be caused if it is applied just before sowing or planting. The best time to apply lime is prior to autumn digging so that crop damage is avoided and the full effect of the liming achieved. It should be applied in the rotational sequence before the lime-loving crops.

In most situations, the natural trend is for soil to become acid so liming is normally required, but there are a few soils where it may be desirable to make them more acid and this presents greater difficulty. However, the incorporation of acid peat into the soil can assist the process. Sulphur and ferrous sulphate are also used with success on some soils (4 oz. per square yard; 135 g. per square metre). If a gardener has a very alkaline soil, he is perhaps best advised to avoid plants which require acid conditions and to avoid using fertilizers such as 'Nitrochalk', which contain calcium.

FERTILIZERS AND MANURES

Fertilizers and manures are applied to the soil to ensure that the plant has sufficient nutrients to enable it to grow satisfactorily. On visiting a local garden shop, a gardener is often faced with a wide variety of types and proprietary brands of fertilizer and this can be confusing. A knowledge of the basic forms of fertilizer can often help in making the correct choice for the vegetables to be grown.

Most 'artificial' or inorganic chemical fertilizers contain one or more of the three major nutrients: nitrogen, phosphorus, and potassium. These nutrients occur widely, but plants can only absorb them in specific chemical forms. Thus, the value of a fertilizer is measured by how much of its nitrogen, phosphorus, or potassium is actually available to plants. The contents of major nutrients have to be given on the label or bag of inorganic fertilizer as '% N' for nitrogen, '% P_2O_5' for phosphorus, and '% K_2O' for potassium. The higher the percentage of a particular nutrient, the greater is the amount of that nutrient in a given weight of a fertilizer. For example, a fertilizer containing 20 per cent P_2O_5 has twice the concentration of phosphorus as one with 10 per cent P_2O_5.

Nitrogen fertilizers

Nitrogen fertilizers are generally in the form of ammonium salts, nitrate, or urea. The common forms include:

	%N
Sulphate of ammonia	21
'Nitrochalk'	21
Ammonium nitrate	34·5
Urea	45
Nitrate of soda	15·5
Calcium nitrate	15·5

Sulphate of ammonia is a common and easily obtainable nitrogen fertilizer which is not readily leached from the soil since the nitrogen in the form of ammonium is held on the clay particles. The ammonium nitrogen displaces calcium from the clay and forms soluble calcium sulphate which is leached from the soil. Since calcium is lost, the effect is to make the soil more acid and, whilst this may be an advantage for acid-loving plants and in calcareous soils, it is generally considered to be undesirable in the vegetable garden. In very sandy soils, which do not contain many clay particles, the ammonium nitrogen will not be retained in the soil and will be leached more quickly. Sulphate of ammonia is a quick-acting fertilizer but not as quick-acting as those containing the nitrate form of nitrogen. It should be applied to moist soil and watered-in if at all possible. The effect of watering-in is to overcome its tendency to scorch leaves and roots, which can cause a check to growth before beneficial effects can be obtained. It is a cheap and easily stored material and its best use is to give a quick boost to growth for leafy crops such as spinach and cabbages.

'Nitrochalk' and ammonium nitrate both contain ammonium and nitrate forms of nitrogen. 'Nitrochalk' also contains chalk which counteracts the acidifying effects of ammonium nitrogen. Both are rapid in their action due to their nitrate content. They are incorporated in the soil prior to sowing or planting as a base dressing and used later as a side dressing for boosting crop growth when extra nitrogen is required. Both of these fertilizers take up moisture, which leads to deterioration and difficulty in handling. It is therefore important that they are stored in sealed plastic bags.

Urea contains a high proportion of nitrogen which is converted to ammonium carbonate in the soil. Gaseous ammonia is formed and can be lost if it has not been well mixed into the soil. Nitrate of soda (Chilean nitrate) and calcium nitrate (nitrate of lime) are rapidly acting nitrogen fertilizers. The fact that plants respond more quickly to nitrogen in the nitrate form make them very useful for side dressings on growing crops. Beet and celery also respond to sodium and it is with these crops that nitrate of soda has its main use. Again, these nitrates absorb moisture and must be kept in airtight containers.

Phosphorus fertilizers

The main forms of inorganic phosphorus fertilizers are superphosphate (20 per cent P_2O_5) and triple superphosphate

(44 per cent P_2O_5). These consist mainly of calcium phosphates which are to a large extent soluble in water but, in the soil, are quickly rendered relatively insoluble and, therefore, unavailable. The phosphate is subsequently slowly released into the soil solution and so becomes available to the plant.

Potassium fertilizers

The two main potassium fertilizers are potassium chloride (muriate of potash, 60 per cent K_2O) and potassium sulphate (sulphate of potash, 49 per cent K_2O). The sulphate is more expensive but is easier to store and is probably the safest to use in the garden as damage can arise from the use of the chloride, particularly in the early stages of growth.

Compound fertilizers

Ready-mixed or compound fertilizers contain varying proportions of nitrogen, phosphorus, and potassium. These are often convenient to use as they are time-saving and are often in a granular form which makes handling easier. A method of abbreviation is often used to describe how much of each available nutrient is in any fertilizer. For example, if a fertilizer contains 20 per cent N, 10 per cent P_2O_5, and 10 per cent K_2O, it is referred to as a 20 : 10 : 10 fertilizer. There are a wide range of compound fertilizers available to gardeners, probably the most common being one which contains equal amounts of N, P_2O_5, and K_2O, for example 'National Growmore' (7 : 7 : 7).

Although the more concentrated fertilizers need less storage space for a given amount of nutrients, they are more likely to cause damage to young seedlings unless carefully applied. With care, however, they can be used successfully and often the gardener can obtain surprisingly good value for money by choosing according to the nutrient contents of the various compound fertilizers available to him.

Organic manures and composts

High prices are often paid by gardeners for organic fertilizers in the belief that they have properties which provide better sources of nutrients than the so-called 'artificial' fertilizers. Whilst the bulky manures do contribute substantially to improving the physical condition of poorly structured soils, so giving better growth, it is equally true that organic manures such as blood, bone or fish meal, composts, farmyard manure, seaweed, and the like depend for nutrient value on their content of nitrogen, phosphorus, and potassium in exactly the same way as chemical fertilizers.

Research has shown that organic sources of nitrogen such as blood and bone or fish meal produce no greater yields than the equivalent amounts of inorganic fertilizers. The claim that they can be used more safely, without damage or scorch to crops, because the nitrogen is released more slowly is open to question since, when finely ground, these fertilizers can release inorganic nitrogen rapidly.

Nevertheless, these materials may be readily available to the gardener and an indication of their nitrogen, phosphorus, and potassium contents are as follows, though it should be noted that they vary according to source and processing:

	%N	%P_2O_5	%K_2O
Dried blood	12–14	2·5	1
Hoof and horn	14	2·5	—
Bone meal	4	7	—
Fish meal	9	2·5	—
Seaweed (dried)	1·5	0·5	5
Wood ashes	—	2·5	5

Farmyard manure varies widely in nutrient content but on average somewhat less than $\frac{1}{4}$ per cent N, $\frac{1}{4}$ per cent P_2O_5, and $\frac{1}{2}$ per cent K_2O may be expected to be available. Thus large amounts are required to supply the plants nutrient needs and much of the benefit achieved is from the improvement of soil structure. Sewage sludges can provide useful quantities of nitrogen and phosphorus but often contain high amounts of metals, particularly when they come from industrial areas. These metals can be toxic to plants so sludge should be used with the utmost caution.

One of the most common and deep-rooted beliefs among certain gardeners and consumers is that crops grown with organic manures are in some way more nutritious than crops grown with chemical fertilizers. There is no scientific evidence to support this view. The plant uses nutrients from the soil in their simplest form and there is no evidence to show that it makes any difference where they come from.

Liquid fertilizers

There are many brands of liquid fertilizer on the market containing varying proportions of the three major nutrient elements and often minor elements. Usually they are sold in solid form which needs dissolving in water or as concentrates which need diluting. Liquid feeding is particularly useful in glasshouses but can also be useful in the vegetable garden when individual plants can be treated for deficiency of a particular nutrient. As the nutrients used for this form of feeding are already in solution, they clearly act more quickly than those applied as solid fertilizers. The instructions on the label of the product being used must be followed to get the best results, since over-use can lead to damage to plants.

Plants also have the ability to absorb nutrients in solution through the leaf as well as through the root so that foliar feeding is possible. The amounts of nutrients that plants are able to absorb in this way are relatively small and invariably it is more worth while to concentrate on feeding through the soil. However, the problem of a soil with a low magnesium content will often take several years to correct by incorporating magnesium-containing fertilizers into the soil, whereas spraying with a 2 per cent solution of Epsom salts (magnesium sulphate) on two or three occasions during the growth of a crop is invariably successful. The minor elements boron, molybdenum, manganese, zinc, and copper can all be applied as foliar sprays to supply these nutrients to plants or, in some cases, to correct disorders.

FERTILIZING VEGETABLES

If all soil was ideally fertile and contained sufficient quantities of all the nutrients to supply the needs for plant growth, there would be no need to apply additional nutrients in the form of fertilizers or manures. Of course this is rarely the case, although some soils have adequate supplies of some nutrients so they do not require further additions. The majority do require regular dressings of fertilizers to ensure optimum yields.

Response to fertilizers

The way in which a plant responds to applications of a fertilizer depends principally on the nutrient content of the soil and the type of crop. Even when no fertilizer is applied, a yield is obtained which represents the effect of the nutrient that is already in the soil. An application of fertilizer may increase the yield substantially, but the application of twice the amount will not necessarily double the increase in yield. The return in extra yield becomes progressively less with each additional portion of fertilizer applied and a point is reached where the cost of applying a larger amount would not be justified for the small return that would be obtained. This type of response is considered to follow 'the law of diminishing returns'.

Obviously less fertilizer will be required for optimum yield if the soil contains large amounts of a given nutrient than if it contains relatively low quantities. So in order to get an idea of how much fertilizer to apply a commercial grower submits samples of his soil to a laboratory and obtains a chemical analysis. By shaking the soil with sodium bicarbonate and ammonium nitrate solutions, the laboratory is able to determine the amounts of phosphorus and potassium in the soil which are 'available' for plants and which the roots can extract. It is not possible to get a measure of nitrogen availability from laboratory tests and the grower uses his knowledge of past cropping on a particular site to help him estimate the amount of nitrogen likely to be available in the soil. Account is also taken of the rainfall, the soil texture, and previous applications of organic manures.

Research over many years has provided much information on the responsiveness of crops to fertilizers and has shown that the response varies depending on the levels of nutrient in the soil and the different vegetables. For example, Brussels sprouts can be grown satisfactorily on a soil containing relatively low amounts of available phosphorus and so require only small additions of this nutrient, whereas lettuce require larger amounts of phosphate fertilizer in order to produce high yields. This provides a good example to dispel the myth that a heavier crop removing more nutrient from the soil must require a larger quantity of nutrient applied as fertilizer, for Brussels sprouts take up far more phosphorus than lettuce but obtain it by more efficiently extracting it from the soil.

Having obtained an analysis of his soil for phosphorus and potassium and having assessed the likely nitrogen status of the soil, the commercial grower will consult tables of responses and get guidance as to the amount of these three nutrients that are necessary for the crop that he is growing. Unfortunately the gardener is not in a position to obtain a chemical analysis of his soil and, unlike the soil test-kits for pH measurement, test-kits for phosphorus and potassium on the market at the time of writing are not altogether reliable. However, 'rapid-tests' have been developed recently at the National Vegetable Research Station and it is hoped that these will be on the market in the near future. The gardener will then be in a position to assess his requirements more accurately. In the meantime, it is possible for gardeners to use the information available on vegetable fertilizer requirements by making certain assumptions which are described later.

Injury to plants by fertilizers

Sometimes the responses to fertilizers are less than expected, particularly when nitrogen is applied. These are invariably caused by an adverse effect of applying fertilizer which offsets the yield-increasing effect. As the amount of fertilizer applied is increased the beneficial effect is dominant at first, producing increases in yield, but as the amounts increase, the adverse effect becomes important and with high levels of fertilizer yields are actually reduced. The effect can be such that with some crops too much fertilizer can produce lower yields than applying no fertilizer at all.

The injurious effects of applying fertilizer are caused by too high a concentration of salts in the soil solution, particularly when nitrates and chlorides are used, causing damage to young roots. Broadcast applications of nitrogen fertilizers before sowing can severely depress the numbers of seedlings emerging as well as causing reduced growth of those that do survive. The effect is most marked in the early stages of growth and, although well-established plants can tolerate the conditions, the potential yield of the crop is hardly ever achieved. Similar effects can be obtained with potassium, especially if muriate of potash is used, but phosphate fertilizers do not produce such a serious problem.

How to apply fertilizers

The adverse effects of applying nitrogen and potassium fertilizer before sowing or transplanting young plants may be overcome in several ways:

Use only part of the required amount of nitrogen fertilizer before sowing and side-dress the remainder after the crop is established.

Spread the nitrogen fertilizer between the rows so that the young roots do not come into contact with high salt concentrations.

Keep the soil moist until seedling emergence is complete or a transplanted crop is established.

Apply potassium fertilizers in the autumn before digging.

Use sulphate of potash which causes less damage than muriate of potash.

Whilst it is generally inadvisable to have nitrogen and potassium fertilizers concentrated near the seeds or young roots, there can be a distinct advantage in placing phosphate fertilizers in this position. Young plants can benefit considerably from extra phosphate supplies, and as phosphate fertilizers tend to become rapidly unavailable to plants in the soil, this provides a ready source of supply in the region which the roots quickly penetrate. Putting the required amount of superphosphate in a band about 2 in. (5 cm.) to the side and 2–3 in. (5–7·5 cm.) below the seed is often found to be more beneficial than broadcast applications since the fertilizer in the band is in less contact with the soil and is not absorbed, so remaining available to the roots for a longer time. Crops responsive to the placement of phosphate fertilizer include onions, lettuces, carrots and French beans.

When to apply fertilizer

Different crops have varying lengths of growing season and common sense tells us that crops which have a short duration of growth, such as radishes and lettuces, need ample nutrients from the start of their growth. In contrast, crops such as Brussels sprouts and leeks, which have a long period of growth, require their nutrients over a considerably longer period of time. Phosphate and potassium can be applied at almost any time prior to sowing or planting, but if a fertilizer containing nitrogen is being used it should be applied just before sowing and/or during the growth of the crop, especially on light soils.

The timing of nitrogen applications is the key to successful growing of many crops. The best advice that can be given is to make sure by top dressing that a crop has ample nitrogen when it is growing most rapidly. If heavy rainfall occurs at any time, it is more than likely that a considerable proportion of the applied nitrogen has been lost from the top-soil.

Interaction between nutrients

The balance of nutrients available to the plant is important and chemical reaction in the soil and at the plant roots can lead to one nutrient element affecting the uptake of another either by increasing (enhancement) or reducing (antagonism) the amount taken up by the plant. The use of a nitrate fertilizer tends to enhance the uptake of potassium from the soil and so it is preferable to use a nitrogen fertilizer containing nitrate when growing a crop which requires a high level of potassium for good growth. Antagonistic effects are far better known, one of the most common being the reduction in uptake of calcium and magnesium by the use of potassium fertilizers. Leaves showing magnesium deficiency are probably those most commonly brought to the scientist or adviser for diagnosis. It probably occurs more on tomatoes than any other crop since they are invariably given high potassium feeds in glasshouses. For this reason, magnesium sulphate (Epsom salts) is often included in feeds for tomatoes.

Many antagonistic interactions between both major and minor nutrients have been demonstrated and a few of the important ones are as follows:

Potassium reduces the uptake of manganese, copper, zinc, and magnesium.
Magnesium reduces the uptake of potassium and calcium.
Nitrate nitrogen reduces the uptake of phosphorus.
Calcium reduces the uptake of potassium and magnesium.

Fortunately, most plants can grow satisfactorily in soils containing widely varying amounts of most nutrients without the risk of serious effects from interactions. Nevertheless, when a deficiency of a particular nutrient occurs (p. 82) it is often helpful to consider whether the problem has been caused by excessive manuring with fertilizers containing antagonistic nutrients.

A SYSTEM FOR THE GARDEN

We have seen how different crops vary greatly in their need for nitrogen, phosphate, and potassium fertilizer, and how soils also vary in their nutrient content. The amount of each nutrient required can be predicted accurately only if the amount already in the soil has been measured by chemical analysis. As the gardener is unlikely to be able to get his soil analysed readily, a method of forecasting fertilizer needs without soil analysis can be used. Although it can only be approximate as it depends on a number of assumptions (in particular, that most plant waste is returned to the soil as compost) it is nevertheless a good guide.

Nitrogen fertilizers behave differently from either potassium or phosphate in soil because nitrogen is readily leached out of the soil by heavy rainfall and fresh applications are necessary for each crop. Furthermore, quite large amounts of nitrogen are required for high yields of many vegetables. On the other hand, the benefits from phosphate and potassium fertilizer usually last for several years and so, in the garden, it does not matter greatly how or when they are applied, or if the amounts vary from year to year. However, the average application over a number of years must be sufficient to maintain levels in the soil above the minimum for good growth of all crops and to replace that taken out of the system by the vegetables we eat. It follows then, that the best plan for using fertilizers in the garden is to adjust the nitrogen application for each crop and to maintain phosphate and potassium at roughly the same levels in the soil for all crops.

Nitrogen

Fertilizers containing nitrogen are usually worked into the soil immediately before sowing, or applied partly at this time and partly as a mid-season top-dressing between the plants. The total amounts of a range of nitrogen fertilizers which can be applied to meet the needs of various vegetables are given in Table 3.1. To find how much fertilizer to apply, first look on the bag and check the percentage N in the fertilizer. Then find, in Table 3.1, the percentage N which is closest to it and apply the amount appropriate for the crop you intend to grow. For example, broad beans will need 4 oz. per square yard (136 g. per square metre) of a fertilizer containing 5 per cent N but only $1\frac{1}{2}$ oz per square yard (51 g. per square metre) of one containing 15 per cent N.

Phosphate and potassium

If your nitrogen fertilizer also contains about the same percent-

Table 3.1

Amount of fertilizer (oz. per square yard) required to meet the nitrogen demands of vegetable crops (to convert to g. per square metre multiply by 34)

	Percentage nitrogen in fertilizer (quoted on bag or packet)						
	5	10	15	20	25	30	35
Peas	0	0	0	0	0	0	0
Carrots; radishes	$1\frac{1}{2}$	$\frac{3}{4}$	$\frac{1}{2}$	$\frac{1}{2}$	$\frac{1}{4}$	$\frac{1}{4}$	$\frac{1}{4}$
Broad beans; parsnips; swedes	4	2	$1\frac{1}{2}$	1	1	$\frac{3}{4}$	$\frac{3}{4}$
Lettuce; onions	$5\frac{1}{2}$	3	2	$1\frac{1}{2}$	$1\frac{1}{4}$	1	1
Calabrese; French beans; turnips	$6\frac{1}{2}$	$3\frac{1}{2}$	2	$1\frac{3}{4}$	$1\frac{1}{2}$	$1\frac{1}{4}$	1
Leeks; potatoes (early)	9*	$4\frac{1}{2}$*	3*	$2\frac{1}{4}$*	$1\frac{3}{4}$*	$1\frac{1}{2}$*	$1\frac{1}{4}$*
Potatoes (main crop); red-beet; spinach; summer cauliflower	11*	$5\frac{1}{2}$*	4*	3*	$2\frac{1}{4}$*	2*	$1\frac{1}{4}$*
Brussels sprouts; summer cabbage; winter cabbage	15*	$7\frac{1}{2}$*	5*	4*	3*	$2\frac{1}{2}$*	$2\frac{1}{4}$*

* For drilled crops, the amount of fertilizer quoted for lettuces should be applied prior to sowing and the remainder after the crop is established.

age of phosphate (P_2O_5) and potassium (K_2O) as of N (e.g. 7:7:7 or 15:15:15) you will automatically apply enough phosphate and potassium along with the nitrogen. If your nitrogen fertilizer does not contain phosphate or potassium, you must apply these two nutrients separately. This can be done, for example, with 2 oz. per square yard (68 g. per square metre) of superphosphate and 1 oz. per square yard (34 g. per square metre) of sulphate of potash either in the autumn or before sowing, but other phosphate or potassium fertilizers could also be used with amounts adjusted according to the percentage P_2O_5 or percentage K_2O. It can also be done by applying farmyard manure at about 10 lb. per square yard (5·4 kg. per square metre).

IF THINGS GO WRONG!

If plants have an insufficient supply of a major or minor nutrient, they invariably show visual symptoms of the deficiency. However, it is often possible to correct a deficiency by treatment of the plants or the surrounding soil. Some of the more common symptoms are now tabulated together with possible remedies. These will aid you in making your own diagnoses and cure.

Major element deficiencies

	Symptoms	**Treatment**
Nitrogen	Usually the first symptoms are pale yellow leaves at the base of the plant where leaves also die prematurely. These leaves have reddish and orange tints in the case of brassicas. Growth is poor	Apply a top-dressing of nitrogen fertilizer around the plants and water in

	Symptoms	Treatment
Phosphorus	In some plants no symptoms appear but growth is poor and slow (lettuce). Lower leaves can exhibit a dull, blue-green coloration	Cure is difficult with the current crop as phosphate fertilizer cannot be watered in. Try placing some superphosphate under the surface of the soil close to the roots
Potassium	Older leaves develop chlorosis (loss of green colour) followed by scorch around the leaf margins (see Fig. 3.2). Brown spots can occur on leaves	Apply muriate of potash around the plants and water thoroughly
Magnesium	Symptoms begin on the older leaves; chlorosis between the veins gives a mottled or 'marbled' effect. The veins remain green and leaves fall from the plant. In tomatoes 'green back' occurs on the fruit (a green shoulder surrounding the calyx)	Apply Epsom salts (magnesium sulphate) as a foliar spray (3 oz. dissolved in a gallon of water (20 g. per litre) every two weeks. May also be watered on to the soil

	Symptoms	Treatment
Calcium	Young leaves 'cup' or curl inwards and the growing point is often deformed or blackened. Calcium deficiency causes 'tip-burn' in lettuces, 'blackheart' in celery, 'blossom-end rot' in tomatoes, and 'internal browning' in Brussels sprouts	Spray fortnightly with a calcium nitrate solution (½ oz. per gallon) (3 g. per litre)

Minor element deficiencies

	Symptoms	Treatment
Zinc	Leaves develop a yellow mottling and young terminal leaflets rosette	Spray with a solution of zinc sulphate (½ oz. per 2 gallons) (1½ g. per litre)
Iron	Young leaves turn pale and chlorotic—almost bleached in severe cases. Veins remain dark green. Occurs mostly on chalky soils	Spray or water with sequestered iron (obtainable at garden shops)
Manganese	Interveinal chlorosis of older leaves giving a mottled appearance. Spots of dead tissue on the leaf ('marsh spot' in peas and beans)	Spray with manganese sulphate solution (½ oz. per 2 gallons) (1½ g. per litre)

	Symptoms	Treatment
Boron	Very similar to calcium deficiency. Terminal buds die. Blackening of crown of beetroot. Hollow or split stems in brassicas	Spray with borax solution (1 oz. per gallon) (6 g. per litre)
Copper	Symptoms vary according to crop. Youngest leaf tips of onions turn yellow/white, wilt, and twist. Young leaves of beans and peas turn greyish green. Only common on peaty soils	Spray with Bordeaux powder or cuprous oxide ($\frac{1}{2}$ oz. per 2 gallons of water) ($1\frac{1}{2}$ g. per litre)
Molybdenum	Leaf lamina fails to develop, resulting in a strap-like leaf. Interveinal yellow-green mottling of older leaves, 'Whiptail' of cauliflower and broccoli—growing point stunted or blind (see Fig. 3.1)	It is too late when symptoms develop! (Water the seedbed with a solution containing 1 oz. ammonium or sodium molybdate per gallon (6 g. per litre) next time this crop is grown)

The symptoms given for the various nutrients are given only as a guide. It should be emphasized that variations occur between crops. Other factors which may give symptoms can resemble or be associated with nutrient disorders. Damage caused by weedkillers and root damage by pests which can cause reduction of nutrient uptake (for example, mild cabbage root fly damage can give the appearance of nitrogen deficiency) and some virus infection symptoms can be very similar to nutrient disorders. Lettuce mosaic virus

Fig. 3.2. Leaves of turnip showing chlorosis around the edges of leaves caused by potassium deficiency.

symptoms are similar to those of magnesium deficiency, both giving a distinct 'marbling' or mottled effect to the leaves.

The remedies outlined are 'last resort' actions but do not exceed the amounts stated. Prevention is of course better than cure and tell-tale symptoms one year should be remembered and noted to ensure that action is taken to prevent a recurrence in future years. Fertilizers are readily obtainable which contain several or all of the various trace elements and can be used if deficiency symptoms regularly appear.

4 Watering vegetable crops

Gardeners are well aware that watering their vegetable crops gives better growth, better quality, and higher yields. This has led to the belief that all vegetables will always benefit from watering, and the more water that is given the better. *This is not so.* It is very easy to give too much water too often.

Apart from the waste of water, time, and effort, watering unnecessarily may merely increase the growth of the plant without increasing the size of the edible part. It may discourage root growth (making the plants more drought-susceptible), it may wash nitrogenous fertilizers out of reach of the roots, and it may reduce flavour. So it is worth understanding how vegetable crops respond to watering. It is also worth knowing how to water different vegetables effectively and efficiently.

How do we decide when water should be given and how much to apply? Fortunately a lot of research has been done on this subject and these questions can be answered for many crops.

However, before dealing with individual crops an understanding of why plants require water, and of the many factors which affect the plants' needs and response to watering, will give a better appreciation of the water requirements of different crops.

WHY PLANTS REQUIRE WATER

Plants take up water from the soil through their roots and it is lost from their leaves into the surrounding atmosphere through small pores or stomata. The loss of water from the plant into the atmosphere is a physical process which is determined by the amount of sunshine, temperature, relative humidity of the air, and also the amount of wind experienced, and, as such, can be calculated or measured. However, the plant is able to restrict its water loss when there is a shortage of soil moisture or when the rate of loss is extremely high—a necessary survival mechanism.

So, during long, hot, sunny days in summer any vegetable crop where leaves fully cover the ground, and which has a plentiful supply of soil moisture, will lose by evapotranspiration more than 1 gallon of water per square yard (5·4 litres per square metre) of crop per day. Conversely, under cloudier, cooler conditions and with shorter days in spring and autumn the water loss per square yard of crop may be as little as 1–2 pints per square yard per day (0·7–1·4 litres per square metre).

Whatever the conditions large quantities of water pass through the plant during the course of its growth, but only a small proportion becomes incorporated into the cell tissue. This

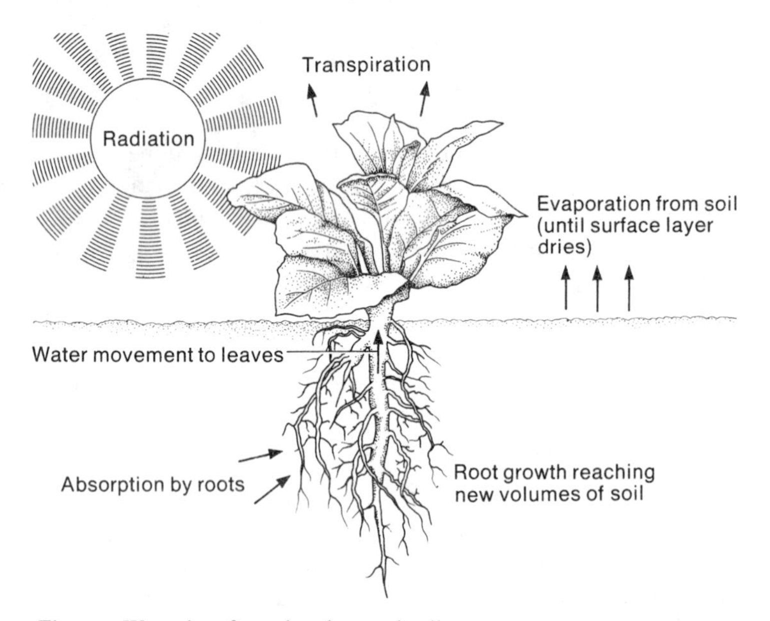

Fig. 4.1. Water loss from the plant and soil.

large throughput gives several benefits: the stomata remain open and all parts of the plant tend to be in a fully turgid condition. This enables photosynthesis, respiration, and other essential processes to take place efficiently. It helps the absorption of nutrients by the roots and their movement within the plant. In addition, the loss of large quantities of water by evaporation from the leaf surface has a beneficial cooling effect under hot conditions.

PLANT RESPONSE TO WATERING

Very few gardeners are curious enough to leave a few plants unwatered to see what difference watering will bring about. But there are often situations when watering will not produce any effect at all and, in some instances, where it will cause unwanted effects. So it is useful to outline the circumstances and factors which cause a plant to respond or be unresponsive to watering and the nature of any response.

Will the plant respond?

If the plant can obtain sufficient water from the soil to sustain the water loss from the leaves at a maximum rate, then watering is unlikely to benefit the growing plant. This is often the situation in the early part of a plant's life when actively growing roots extend rapidly into new areas of wet soil where they can absorb water generally without restriction. But if rainfall is inadequate to replenish dry soil, if the soil cannot hold a plentiful supply of water, if there is competition for water from neighbouring plants, or if root growth is insufficient to tap new water reserves or is restricted in any way, then, under these conditions, plants will respond to watering.

It is worth looking at these different aspects in a little more detail. However, it must be remembered that vegetables are growing in the garden environment over which we have little control and that we are considering a dynamic, changing situation where one factor affects the plant's response to another.

The soil water reservoir

The amount of water held in the soil reservoir is obviously a crucial factor affecting any plant's response to watering. Water is retained in the soil both in the pore spaces and as a thin film over the surface of the particles (see Fig. 4.2). When soil is holding the maximum amount of water under freely draining conditions it is at 'field capacity'. As the soil gradually dries out by evaporation from the surface or by extraction by plant roots, the water is held by the soil with greater and greater surface-tension forces. As a result the plant has greater and greater difficulty in absorbing the water until a stage is reached when the roots cannot take up the water fast enough and the plant starts to wilt. When the plant does not regain its turgid state even overnight then the soil moisture content has reached the 'wilting

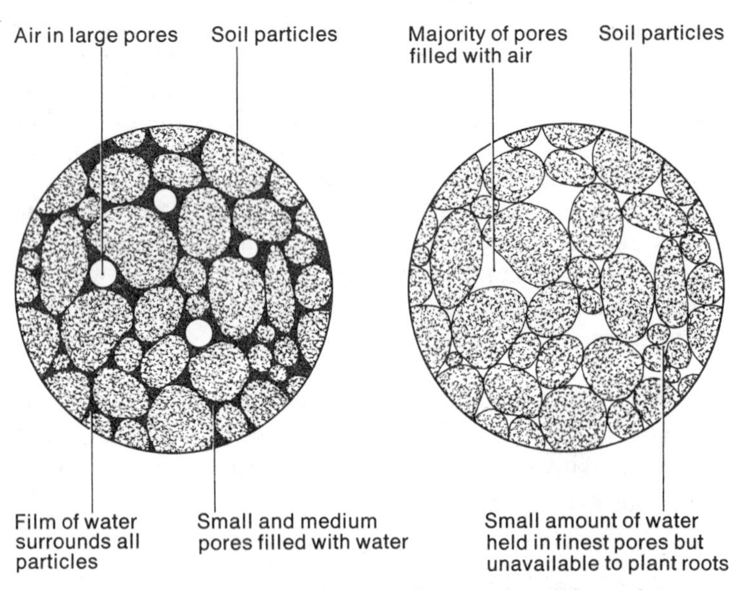

Fig. 4.2. Representation of the soil moisture status at the condition of field capacity and permanent wilting point.

point' and the plant will eventually die if it is not watered. The water held by the soil between these two conditions of field capacity and the wilting point is 'available water' for plants.

Obviously the soils which contain the most available water will sustain the plants for the longest time before wilting occurs. Even at the wilting point there is still water in the soil, but it is held in the smallest pores with such force that the plants cannot absorb it and so is called 'unavailable water'.

The amount of available water held in a soil will depend on the size of the individual particles present, which gives differences in soil *texture*, and on the way the particles are grouped together, which alters the *structure* of the soil.

Soil-texture effects. Observant gardeners know only too well that if the soil contains a high proportion of coarse sand particles then water drains freely through it and very little is retained. On these soils plants need to be watered frequently. But if there is a good mixture of coarse and fine sand together with some smaller particles of silt and clay then these loams are good retentive soils which do not dry out so quickly as the sands. The amounts of available and unavailable water held in the surface foot of representative soils of different textures are shown in Fig. 4.3. It can be seen that a plant growing in the silt loam would have four times the amount of water available to it than a comparable plant growing in the coarse-sand soil; it would therefore not require watering so frequently.

There is not much that even the gardener can do about altering the *texture* of his soil to make it more retentive of water. In days gone by clay was often spread on coarse sandy soils to help water retention, but this is no longer done even on a small scale. However, the gardener can influence the *structure* of his soil quite considerably.

Soil-structure effects. A well-structured soil can hold considerably more available water than a poorly structured soil of the same texture. This is because the structure gives more pores of the right size to hold water which is readily absorbed by plant roots. In a well-structured soil the mineral particles are grouped together with colloidal and organic material to form stable soil crumbs which give a good tilth. Such soils do not get too compact and give an ideal medium in which an optimum combination of water and air and an adequate pore space is provided for root growth. These conditions allow greater and more thorough root penetration and more efficient extraction of the water.

The most effective way of improving soil structure and thus increasing the available water capacity of a soil is to dig in farmyard manure, home-made compost, peat, or other bulky organic manures. These should be added in large amounts, at least 10 lb. per square yard (5·4 kg. per square metre) per year, on poorly structured, coarse-textured soils. The manures should always be well rotted and they should be thoroughly mixed with the soil over the full depth of rooting if they are to be most effective. They should not be buried in a solid mass at the bottom of a trench. The benefits are likely to be substantial. For example, studies showed that after applying 10 lb. per square yard of

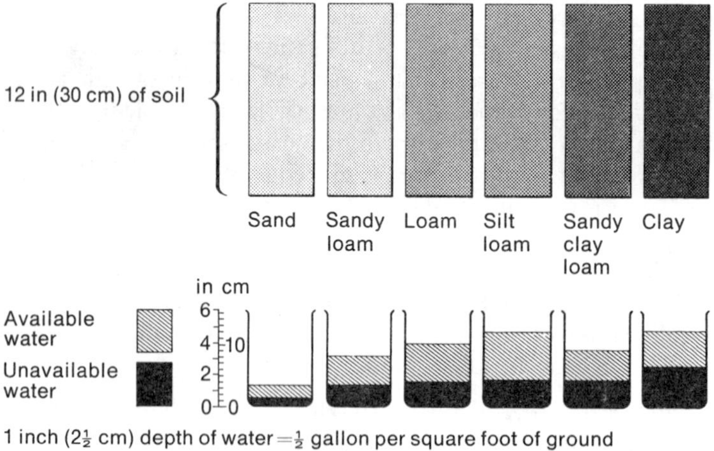

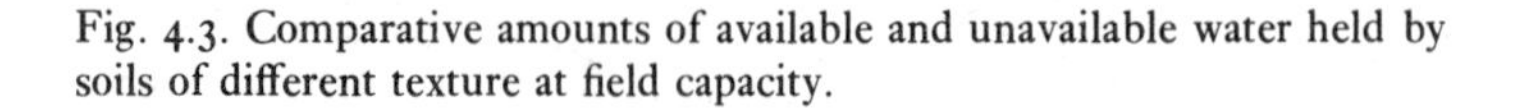

Fig. 4.3. Comparative amounts of available and unavailable water held by soils of different texture at field capacity.

farmyard manure before each crop over a six-year period, the amount of available water held in the top 18 in. (45 cm.) of a sandy loam soil had been increased by over 25 per cent and in

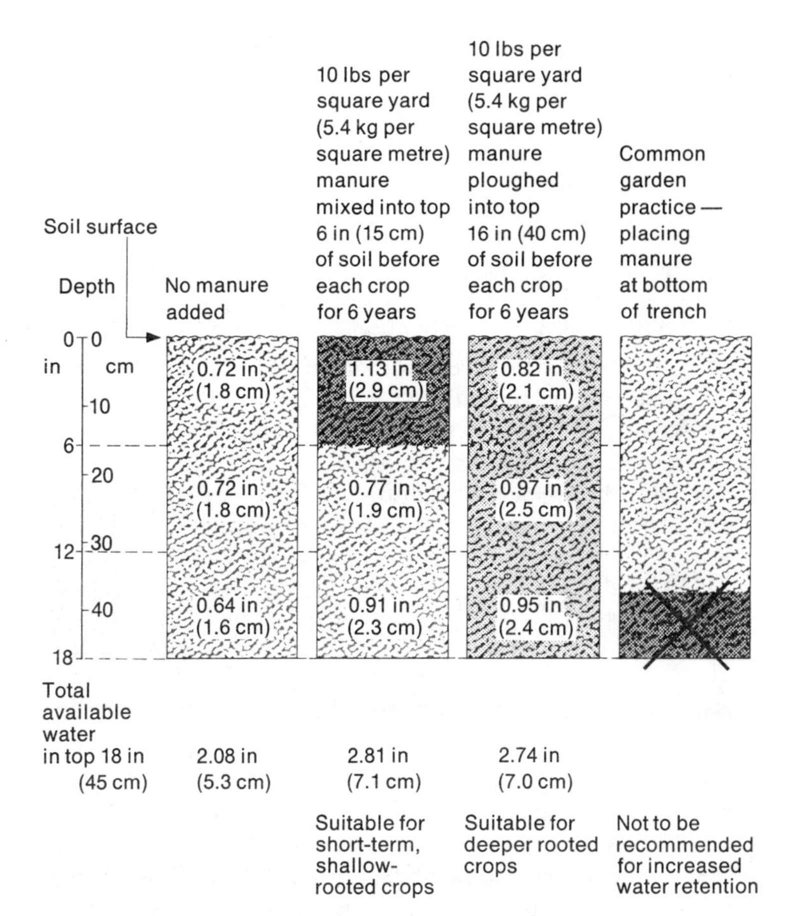

Fig. 4.4. An example of bulky organic manures increasing the amount of available water retained in a sandy loam soil.

the surface 6 in. (15 cm.) of soil by over 70 per cent (see Fig. 4.4). This is one sure way in which the need for watering of crops in the garden can be reduced.

Drying out of the soil

As the soil dries out through plant roots extracting water and evaporation from the surface, a soil moisture deficit will arise. This is normally measured in terms of the depth of water, in inches, which would be required to rewet the soil to the condition of field capacity. So during a settled period of dry, sunny weather, a well-grown crop which completely covers the ground could remove from the soil each day up to 1 gallon of water per square yard (5·4 litres per square metre), increasing the soil moisture deficit by about 0·2 in. (0·5 cm.). Conversely, during unsettled weather with frequent showers and some heavier periods of rain, the soil reservoir is 'topped up' by rain water. The amount may completely replace or exceed the water lost by evapotranspiration over the period and in these latter circumstances the soil moisture deficit would be reduced. Commercially, the average soil moisture deficit can be calculated for any part of the country from meteorological data in order to work out the need for irrigation of different crops.

The gardener knows that, normally, the wetter the soil the better will his crops (and weeds) grow. However, as the soil dries out there may not be much change in the appearance of his plants until the wilting point is reached, but long before this stage plant growth will have been affected as water absorption becomes more difficult for the plants. So, in deciding when to water vegetables in dry periods we have to strike the right balance between waiting until the plants wilt, with a consequent small or big loss of yield, or watering frequently with all the time and effort that that involves. Fortunately definite guidance can be given for the majority of vegetable crops and will be found later in this chapter.

Competition for water

During the early phases of plant growth the roots will grow

rapidly, continually working into fresh areas of soil to absorb water and nutrients. This phase will continue either until root growth is reduced because of limiting factors within the plant often correlated with flowering and fruiting, or because the volume of soil available to the plant has been fully occupied. This volume of soil can be limited in depth by poor soil conditions preventing penetration by the roots or, laterally, by the presence of other plant roots. So if plants are spaced widely apart the roots of each plant have a greater volume of soil from which to extract the available water and, therefore, the larger the plants can grow without the need for watering. Conversely, when plants are grown closely together the volume of water which can be extracted by each plant is restricted. Consequently, the more frequently do they need to be watered if growth is not to be adversely affected.

To illustrate this effect the results of experiments have shown that if plants of early summer cauliflower, which are very sensitive to water shortage, were grown at normal commercial spacings of 24 × 24 in. (61 × 61 cm.) or closer, they responded to irrigation by producing larger heads. If, however, they were grown at a wide spacing of 34 × 34 in. (86 × 86 cm.) the plants did not respond to watering by growing better or producing larger curds, even under drying conditions. Obviously, the root systems of the widely spaced plants had been able to exploit fully the extra water held in the bigger volume of soil. So for certain plants watering may be unnecessary if wider spacings are used. This is another way in which the need to water can be reduced, providing space is not at a premium in the garden.

It must be remembered that weeds strongly compete for the limited quantity of available water held in the soil. They should be removed therefore as soon as they appear or can be handled. In this way all the soil water reserves can be used by the crop plants.

Mulching crops with compost, leaves, or peat is a useful way of preventing water loss from the soil surface. To get the maximum effect the materials should be put around the plants after watering or following rainfall.

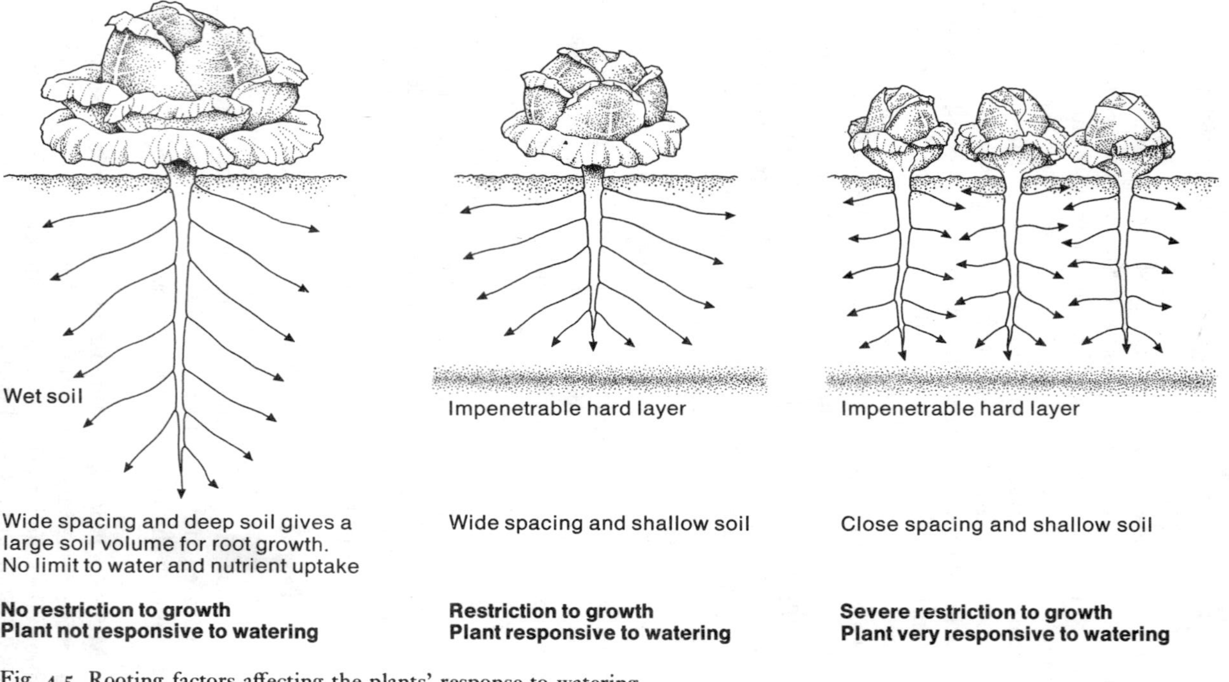

Fig. 4.5. Rooting factors affecting the plants' response to watering.

Root growth

It has been mentioned that root growth may not be maintained at a high rate throughout the life of some crops. In general whenever annual plants pass through the flowering and fruiting phases of development root growth is very much reduced because the plant gives priority to the development of flowers, fruit, and seeds. As a result of this reduced root activity instead of the roots growing into new regions of soil, water absorption becomes dependent on the movement of water through the soil to the root surface. But this water movement in soil cannot take place at a rate sufficient to supply the plant's needs except when the soil is very wet within twenty-four hours of rainfall or watering. Thus these plants become very responsive to soil moisture conditions at flowering and fruiting. They are consequently known as 'moisture-sensitive' stages of growth.

The nature of any response to watering

The situations and factors which affect whether vegetable plants will, or will not, respond to water have been outlined. We can conclude that, except in certain circumstances, whenever the soil dries out to any appreciable extent, say after seven to ten days without rain, then plants will respond to watering in a number of different ways. Some are beneficial as far as the gardener is concerned; some are not. It all depends on what we are trying to produce from the crop. Is the object to produce as much leaf as possible as with lettuce or cabbage? Is it to produce high yields of roots such as carrots, or tubers such as potatoes? Or is it to produce the edible seeds of peas and beans, or fruits such as tomato or marrow? Watering will affect the growth of the different plant organs in different ways and it is useful to understand the basic principles of the growth response in order to choose the appropriate watering strategy for each type of crop.

The nature of the plant response will depend on the type of crop and the stage of growth at which watering is done. Watering generally results in a rapid increase in leaf and shoot growth. So with summer cabbage, for example, increases in size of head at harvest time have been found to be roughly proportional to the amount of water given throughout growth. This effect is to be welcomed because the whole of the cabbage head is edible and what is required.

Root crops such as carrots or red-beet may also respond to frequent watering by increased leaf growth. However, this is not necessarily accompanied by a proportional increase in the edible part of the plant, which in these crops is the root. So with these crops a more judicious use of the watering can is required.

Similarly, too much water available to pea, broad bean, and dwarf French bean plants early in their growth will result in excessive vegetative growth without any beneficial effects, and even adverse effects, on the yields of peas or beans produced. When these crops start to flower and produce pods, however, then watering becomes very beneficial. The reason for the change lies in the nature of the plant's response. At flowering time a plentiful supply of water will increase the number of flowers which form pods and, in the case of peas and broad beans, will increase the numbers of seeds which set in each pod. Watering these crops as the pods are swelling will tend to increase the size of the peas and beans and the number of pods. So with these legume crops the effects of watering will vary at the different stages of development.

Apart from the effects on the weight and size of the edible parts of the plants watering can affect the time of maturity of the produce. Watering during dry periods can so enhance the growth of crops such as early potatoes that earlier lifting can take place. On the other hand watering of bulb onions late in growth may delay the natural ripening of the bulb. So it is not possible to generalize on this aspect.

Watering can also affect the quality and flavour of vegetables. In very general terms the more water that is given to crops the less flavour they have. It then becomes a matter of personal taste whether, for example, a strong or weak flavoured turnip is preferred. As a general guiding rule the conditions which give high yields, large-sized vegetables, and good visual appeal are not usually the same conditions which result in highly flavoured vegetables.

Reducing the need for watering

Points from previous pages of ways in which the summer chore of watering can be reduced may be summarized as follows:

By improving the soil water reserves
—add bulky organic manures and mix thoroughly with the soil.

By encouraging greater ramification of the plant roots
—increase the rooting depth by deeper digging
—space transplanted crops wider apart.

By conserving water in the soil
—mulch the soil surface
—remove all weeds.

HOW TO WATER CROPS MOST EFFECTIVELY

When watering cannot be avoided but water supplies or the time to water are limited then the 'art' of watering needs to become a science so as to become as efficient as possible. The aim should be to obtain the maximum possible yield response from a given quantity of water applied. Research on this subject has indicated three major ways in which water may be more efficiently used in commercial irrigation practice. They are: watering at moisture-sensitive stages of growth, limiting the frequency of watering, and limiting the amount of water given, either singly or in combination. These findings can be equally applied to the garden situation.

Water at moisture-sensitive growth stages

One obvious way of making the most efficient use of limited amounts of water is to irrigate only at those stages of growth when the most beneficial response is obtained and to withold water at the other growth stages when increases in the yield of the edible parts are not obtained. The example of the moisture-sensitive stages of growth of peas and beans, those of flowering and podding, have already been noted and the reasons for the increases in the yield already outlined.

A similar pattern of response is shown by other vegetable crops grown for their seeds and fruits, such as tomato, marrow, and sweet corn. In these crops, too, there is a reduction in root activity during flowering and as the fruits are developing and so the plants become particularly responsive to watering at these times.

With maincrop potatoes nearly all varieties respond with an increase in tuber yields when the crop is watered after the tubers have been formed. A plentiful supply of water during the growth of the tubers gives higher yields and larger potatoes at lifting time.

Limiting the frequency of watering

Some crops do not have moisture-sensitive stages of development and to obtain *maximum* growth and yield it is often necessary to water them as frequently as possible. However, with many of these crops the response to more frequent watering shows that the law of diminishing returns is operating. For example, four waterings may have given a much smaller increase in yield for each watering than a single watering. In these circumstances, therefore, reducing the number of waterings should be a more efficient way of watering.

To illustrate this point the results of experiments with summer cabbage grown on a sandy loam soil may be quoted. This crop is very sensitive to water shortage at all stages of growth so that to get the maximum rate of growth and the maximum size of head the plants must be watered frequently. It was found that when the crop was watered eleven times between planting and harvest the head weight was increased by 100 per cent compared with plants receiving only rainfall. However, if only two waterings were given the weight was increased by 80 per cent whilst a single watering, given about two weeks before the cabbage was ready to cut, increased the head weight by 65 per cent.

Obviously, in terms of the effort put into applying the water, the most efficient strategy was to water once only, a fortnight before they were ready for cutting.

This general strategy of watering on fewer occasions if water or time is short can be applied to most of the crops where the leaves and shoots are eaten, for example the brassicas, lettuce, and celery. The most effective time to give a single watering to all these crops is between ten to twenty days before they are expected to mature. In a similar way if watering of potatoes has to be limited a single application should be given just after the tubers have formed when they are at the 'marble' stage (see p. 109).

It should be emphasized that this limited watering policy will not produce maximum growth and yields of these crops but will be one of the most efficient methods of watering.

Limiting the quantity of water applied

The third method of maximizing the desired response for the minimum of effort is to give smaller quantities of water to the crops than is necessary to rewet the whole depth of the soil. This is *not* saying: give a little and often. It is indicating that if watering is done less frequently, following the advice given in the last section, then it is generally unnecessary to thoroughly soak the soil below a depth of 1 ft. For example, it may be decided that a summer cabbage crop should not be watered until two weeks before cutting. Depending on the weather the soil may have dried out to such an extent that it would require up to 12 gallons per square yard (65 litres per square metre) to rewet it down to the full rooting depth of the crop. By only applying 4 gallons per square yard (22 litres per square metre) and, therefore, only wetting perhaps the surface foot of soil, a maximum response to this quantity of water will be obtained.

With the exception of watering seedlings and transplants it is not really worth while giving less than 2 gallons per square yard (11 litres per square metre) at any one time, especially in hot weather. This is especially so if the water is applied evenly over the soil surface because a significant proportion of that quantity would be evaporated from the wet soil surface and be of no use to the plants.

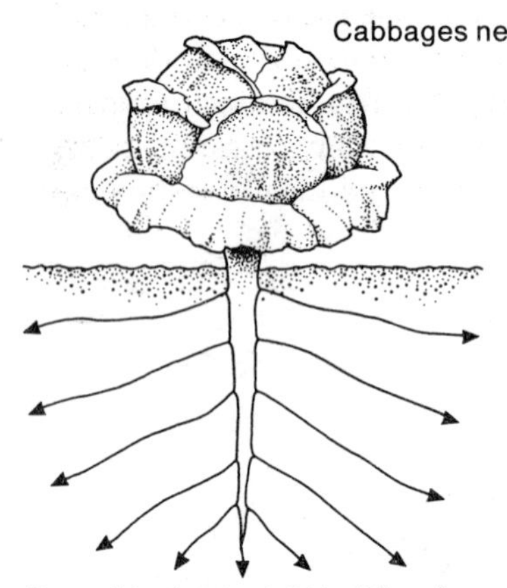

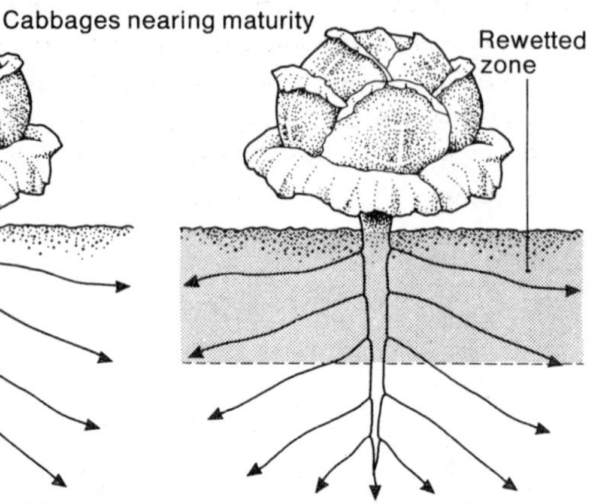

Dry soil to depth of 24 in (61 cm) needing up to 12 gallons per square yard (65 litres per square metre) to return to field capacity

Before watering

4 gallons per square yard (22 litres per square metre) wets the surface 12 in (30 cm) and a maximum response will be obtained

After watering

Fig. 4.6. Efficient watering: limiting the quantity of water given to the crop.

RECOMMENDATIONS FOR WATERING THE DIFFERENT CROPS

All vegetable crops

To aid seed germination. Adequate soil moisture is essential for the establishment of all crops, whether grown from seed or transplants. For crops which are not to be transplanted, but are to be grown to maturity where they are sown, it is essential that soil moisture is adequate to ensure good, uniform, seed germination. The best policy is to try to avoid having to water the soil after sowing as, on some soils, a surface 'cap' or crust will form and prevent the seedlings from emerging (Chapter 2). So, if the soil is dry, dribble water along the bottom of the seed drill immediately before sowing the seed using about 1 gallon of water to 20 ft. (1 litre to 1·3 metres) of row (see Fig. 4.7). Alternatively, if the soil is too dry to produce a good tilth, water the whole area just before making the final seedbed preparations. After one to two days, depending on the soil, a final raking will produce a good seedbed and the seeds can then be sown into moist soil.

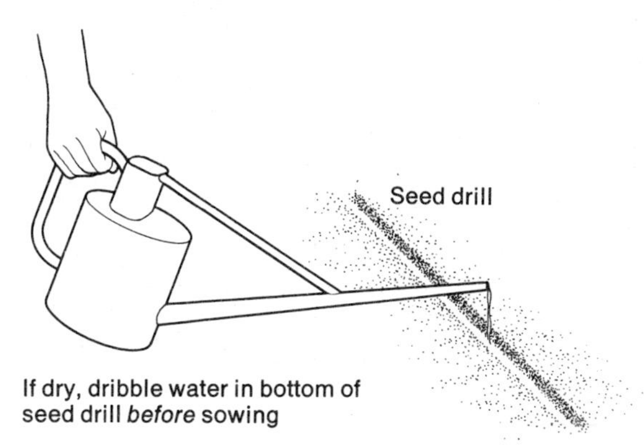

Fig. 4.7. Watering to aid seed germination.

If it does become necessary to water before the seedlings have emerged and also when watering small seedlings, use a fine rose on the watering can or hose to give small droplets. This will do less damage to the plants or to the soil structure. Should a 'cap' form on the surface following heavy rain before the seedlings emerge, keep the soil surface wet by frequent light waterings and the seedlings will be able to push their way through it.

To establish transplants. Transplanting is a traumatic shock for all plants; roots are damaged and the plant cannot replace water lost from its leaves fast enough and wilting occurs. It is, therefore, of crucial importance to water the transplants frequently, daily if possible, with not more than $\frac{1}{4}$ pint (142 ml.) per plant on each occasion. There is no point in applying more if the water is directed around the base of the plant; do not spread it all over the soil surface (see Fig. 4.8). During the recovery period if the weather is sunny the plants can be covered with newspapers to reduce the water loss from their leaves.

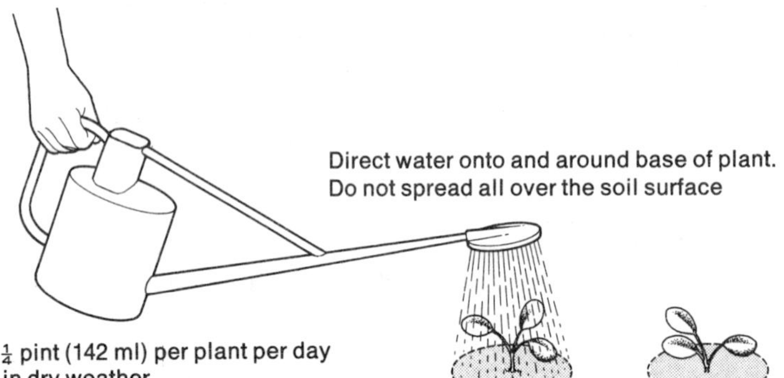

Fig. 4.8. Watering after transplanting.

To water-in fertilizer. For all crops, small quantities of water can be used to wash top-dressings of fertilizers into the soil during dry periods, so making the nutrients available to the plants. It can also be beneficial in the early spring when there is no shortage of water. In these circumstances it is done to make available top-dressings of nitrogenous fertilizers to crops such as spring cabbage which may be short of nitrogen if it has been leached out by winter rains.

Leafy vegetable crops

Those crops where the leaves or the shoots are eaten—for example, cabbage, cauliflower, sprouting broccoli, kale, calabrese, lettuce, celery, and spinach—should be watered frequently to obtain maximum yields. The most tender, succulent crops are produced when the plants have never been checked in their growth either by shortage of water or nutrients. Especially with these crops it should be remembered that the maximum response to watering can only be obtained if there are sufficient nutrients available, and especially nitrogen, to maintain rapid growth.

With celery, daily waterings can be very beneficial under dry conditions. From sowing in the seedbed the aim should be to encourage a fast rate of growth by frequent watering, at least weekly, during dry weather. With 'difficult' crops such as cauliflower, checks to growth due to water shortage at this early stage may have harmful effects much later in life, resulting in premature heading (buttoning) in some varieties when small early curds are produced.

Transplanted crops should be watered regularly until the plants have become established. Thereafter, as a general guide, 2–3 gallons of water per square yard (11–16 litres per square metre) each week will keep the plants growing well in dry summer months. If watering such as this is impossible, wait until two weeks before the crops are ready to be cut and then give 4 gallons per square yard (22 litres per square metre) if both the weather and the soil are dry; this treatment will give a rapid increase in plant weight during the last two weeks of growth. More water than this should not be given to crops such as cabbage and lettuce, otherwise bursting of the heads may occur.

With Brussels sprouts the plants should be watered-in after planting. Thereafter it is unlikely that watering is worth while with this widely-spaced crop except possibly in extremely dry conditions such as those experienced in 1976.

Fruiting vegetable crops

Legumes—peas and beans. Crops of peas, broad beans, and French and runner beans give big increases in yield of the edible seeds and pods if they are watered during flowering and as the pods are growing. Once the seedlings have emerged through the soil they should not be watered, unless severe wilting occurs in very dry weather, until the plants start to flower, otherwise too much stem and leaf growth will be made. Then, throughout the flowering and pod-growing periods, water should be given twice a week at the rate of 1–2 gallons to the square yard (5–11 litres per square metre). The water should be directed to the base of the plants.

If watering of these crops has to be limited for any reason, 1–2 gallons to the square yard should be given as the first flowers start to open. This should be followed by a watering as the pods are swelling. Both waterings will be very beneficial in dry weather.

As we have seen, watering at flowering time will tend to increase the number of peas and broad beans which set within the pods. In addition a plentiful supply of water to the roots as the pods are swelling will increase the size of the individual peas and beans and will delay the time when they go overmature and become tough and 'starchy'. With French and runner beans frequent watering (and frequent picking of the beans) will increase pod-set and yield and will also improve the quality of the pods by delaying the onset of 'stringiness' and maturity.

It is a long-held belief that the problem of pod-setting in runner beans is associated with dryness of the atmosphere around the flowers during flowering. The traditional remedy has been to create humid conditions around the flowers by syringing or spraying the plants once or twice a day during dry weather. In fact, experimental work over several years, including very dry ones, has shown that syringing never had a beneficial effect and often has tended to *decrease* the number of pods which set. It is possible that the yield increases claimed for this practice are the result of actually applying sufficient water so that it ran down the plant to the soil and acted as a light irrigation. Indeed, there is some experimental evidence that small quantities of water applied to the soil during flowering will give better pod-set than syringing the plants.

Other fruiting vegetables. Crops such as tomatoes, marrows, cucumbers, and sweet corn also require plenty of water at flowering time and as the fruits are setting and growing. With these crops the plants should have top priority for any limited quantities of water after transplanting in order to get rapid establishment.

With outdoor tomatoes once growth has re-started after planting watering will not normally be required until flowering starts. This treatment will encourage the growth of the root system. (If plants are grown in bags of peat or compost regular watering will be required throughout growth because of the restricted volume available to the roots.) From flowering-time onwards plenty of water will enable fruits to set and grow but at the same time allow shoot growth to continue, which is essential if high yields are to be obtained. Shortage of water during fruiting results in smaller fruit and smaller total yields. It can also make worse certain disorders such as blossom-end rot which gives a typically black shrunken area at the end of the fruit.

There is some evidence with this crop that fruits produced under dry conditions, although smaller, have much more flavour than those grown under wet conditions. It therefore becomes a matter of judgement for the gardener. If maximum yields of large fruit are required plants should be watered daily in hot sunny weather. If lower yields and smaller, more highly flavoured fruits are wanted the plants should be watered about twice a week, but the fruit should be examined regularly to make sure that the tell-tale symptoms of blossom-end rot are not appearing.

With sweet corn, also, once the plants have got established watering is not normally needed until flowering (tasselling) starts. Then watering at this stage will increase the number of seeds formed on each cob and a plentiful supply as the seeds are filling will increase the size and improve the quality of the cob.

When growing marrows and courgettes plenty of water should be given throughout their life. In the early stages it is necessary to get rapid growth in order to build up a big framework with a large leaf area to sustain the fruits when they form. Once fruiting has started copious amounts of water are needed at frequent intervals as these fruits are nearly all water.

With widely spaced plants such as outdoor tomatoes and

marrows, water is conserved if a small pot is sunk into the ground close to the plant and water is poured only into the pot and not on to the surface of the soil.

Root vegetable crops

With such crops as carrot, parsnip, beetroot, and radish too frequent watering will result in unbalanced, lush, foliage growth. The aim should be to keep the plants growing steadily by watering every two to three weeks if necessary, before the soil gets very dry. In this way root quality will be improved. When watering every fortnight in dry weather 3–4 gallons of water per square yard (16–22 litres per square metre) should be given on each occasion to a crop when the storage roots are actively growing. For a crop at an early stage of growth 1 gallon of water for each yard (5 litres per metre) of row may be sufficient.

Watering (or rain) after a prolonged dry spell will often cause the roots of carrots and parsnips to split; that is why the soil should not be allowed to become too dry.

Potatoes

With this crop different watering policies will be required, depending on whether early harvests or high yields are wanted and also which varieties are grown.

Earlies. If a high yield of an early variety is the objective then the crop should be watered throughout the growing period at ten to fourteen day intervals, giving 3–4 gallons of water per square yard (16–22 litres per square metre) on each occasion. In this way rapid growth is encouraged and if the plants are not lifted until the foliage starts to die-back high yields will be obtained.

However, if earliness is of overriding importance, water should not be given until the tubers have reached the size of a marble (the 'marble' stage, as illustrated in Fig. 4.9). Then one good watering should be given, applying about 4 gallons per square yard (22 litres per square metre).

From the limited experimental evidence at present available

on the response of early potato varieties to different soil moisture conditions it would appear that 'Vanessa', 'Arran Pilot', and 'Home Guard' are fairly tolerant of drought conditions. In contrast, 'Maris Peer' and 'Ulster Sceptre' appear to be susceptible to dry conditions and under such conditions yields can be much affected. On this evidence 'Maris Peer' and 'Ulster Sceptre' would respond much more to watering than the other varieties named.

Maincrop. For maincrop potatoes water should not normally be given until the tubers start to form. This stage coincides approximately with that of flowering and for most varieties a single, heavy watering of 4–5 gallons per square yard (22–27 litres per square metre) when the tubers are at the marble stage

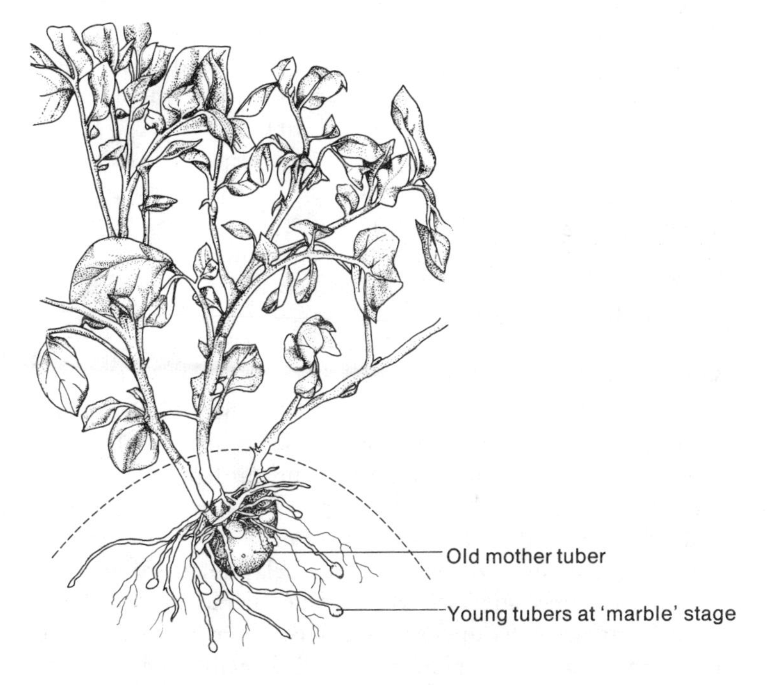

Fig. 4.9. 'Marble' stage of growth when potatoes should be watered.

can be very beneficial. The effect is to increase the size of the tubers and this can be very welcome if the variety tends to produce a large number of small- to medium-sized potatoes. Examples of such varieties are 'King Edward' and 'Ulster Torch'. However, if the variety characteristically produces only a small number of potatoes, for example 'Majestic', then watering should be given ten to fourteen days before the tubers start to form. This will stimulate the production of more tubers per plant. If this is followed by a watering at the marble stage, yields of these varieties will be improved both by an increase in the number and the size of the potatoes.

It is important to try to keep the tubers growing steadily so as to reduce the number of misshapen tubers and those with cracks and secondary growth which result from stop–go conditions for growth.

Again, from limited experimental evidence, it would appear that 'Pentland Crown' shows tolerance to drought conditions and, therefore, would be a suitable variety to grow where watering of the crop would be impossible. On the other hand 'King Edward' and 'Maris Piper' appear to be susceptible to dry conditions, and would be expected to respond well to watering at the marble stage; the variety 'Majestic' appears to be intermediate in its response to soil moisture conditions.

There is evidence that watering reduces the number of tubers affected by common scab.

Onions

Bulb onions. Little experimental work has been done on the water requirements of this crop. As it has sparse foliage, is normally grown at wide spacing in the garden (to get large bulbs), and generally develops a deep root system, it is not considered to respond well to watering. Furthermore, wet conditions at a late stage of bulb growth delay maturity and are believed to affect adversely the storage characteristics of the bulbs. Undoubtedly, under very dry conditions such as those experienced in 1976, onion plants will benefit from frequent watering, especially in the early stages of growth before the plants have developed a good, deep, root system. It would be prudent not to water after the middle of July for the late, keeping varieties.

For the very early, overwintered, Japanese varieties which mature in June and July such reservations are not so important. The bulbs of these varieties will not keep for more than two to three months and so are grown for immediate consumption. Therefore they can be watered right up to harvest time under drought conditions. However, under moderately dry conditions it is unlikely that this overwintered crop will respond to late watering for, by this time, the plants have developed a deep root system.

For this crop it is essential for successful overwintering that the seeds are sown at the correct time in August for the area and that germination is immediate and rapid. At this time of the year the soil is quite likely to be too dry for successful germination of the seed and so watering of the drills before sowing, or the seedbed after sowing, may be essential to get satisfactory establishment.

Salad onions. This crop is sown at intervals throughout the spring and summer to provide a continuous supply of produce. Watering of the seedbed may become essential at certain times during the summer to get satisfactory seed germination. Thereafter plenty of moisture in the soil will encourage rapid growth of this crop.

Seed crops

Many keen gardeners save their own seed from particularly good vegetables they have grown. It is known that root growth of many crops is considerably reduced at flowering time and as the fruits and seeds are forming so water absorption is often adversely affected. The quality and size of seeds produced can also be greatly affected by the weather conditions prevailing at this crucial stage of development. It is, therefore, very beneficial to water at frequent intervals any seed-producing plants at their moisture-sensitive stages. The water should be directed only to the base of the plants.

Table 4.1
A summary guide to watering your vegetable crops

Crop	Category* (see footnote)	Time to water	Amount of water to give in gallons per square yard (multiply by 5·4 to convert to litres per square metre)	Additional comments
Beetroot	C	Before soil becomes too dry and growth is checked	2 (along row in early stages)	Improves the size and quality of the roots
Beans				
broad	B	At the start of flowering and throughout the pod-forming and picking period	4 per week depending on rainfall and weather conditions	Improves pod-set and quality. Syringing of runner beans is not helpful (see p. 106)
dwarf French	B			
runner	B			

Broccoli				
(winter cauliflowers)	C	For three to four weeks after transplanting in July to get established	Up to $\frac{1}{4}$ pint (142 ml.) per plant per day in dry warm weather	Further waterings may be required in August and September if weather dry
sprouting	C			
Brussels sprouts	C	For three to four weeks after transplanting to get established	Up to $\frac{1}{4}$ pint (142 ml.) per plant per day in dry warm weather	Watering not normally required after plants are established.
Cabbages (all types)	A	As frequently as is practicable in dry weather	Up to 4 per week	If water is short give single watering two to three weeks before harvest is expected
Calabrese	A			
Carrots	C	Before soil becomes too dry and growth is checked	2 (along row in early stages)	Watering at later stages after soil has got too dry may cause splitting
Cauliflowers	A	As frequently as is practical throughout growth	Up to 4 per week	If water is short give single watering two to three weeks before harvest is expected

Table 4.1 (*cont.*)

Crop	Category (see footnote)	Time to water	Amount of water to give in gallons per square yard (multiply by 5·4 to convert to litres per square metre)	Additional comments
Celery				
self blanching	A	As frequently as is practical throughout growth	Up to 4 per week	Improves the size of stick and its quality
trench	A			
Leeks	C	Water after transplanting to get established	$\frac{1}{8}$ pint (70 ml.) per plant per day in dry warm weather	Additional waterings can be given if large leeks are required
Lettuces	A	As frequently as is practical throughout growth	Up to 4 per week	Improves size and quality. When water is short give application seven to ten days before harvest

Marrows (and courgettes)	A	As frequently as is practical throughout growth	May need 2 twice per week	Water copiously as soon as fruits form
Onions				
bulb	C	Water if necessary to get plants established	Up to $\frac{1}{8}$ pint (70 ml.) per day if dry	Do not water in later stages of bulb growth
salad	C	Water before sowing to get seed germination if dry	1 gallon to 20 ft. of row (1 litre to approx. 1·3 metres of row)	Can be watered at any stage if soil very dry
Parsnips	C	Before soil becomes too dry and growth is checked	2 (along row in early stages)	Improves size and quality of roots
Peas	B	At start of flowering and throughout the pod-forming and picking period	4 per week depending on weather conditions	Improves pod-set and quality
Potatoes				
early	B	For earliest crops water at 'marble' stage	4	Response depends on variety (see p. 109)
early	A	For high yields water throughout growth every ten to fourteen days	3–4 on each occasion	Response depends on variety (see p. 109)

Table 4.1 (*cont.*)

Crop	Category	Time to water	Amount of water to give in gallons per square yard (multiply by 5·4 to convert to litres per square metre)	Additional comments
Potatoes (*cont.*)				
maincrop	B	For most varieties water at 'marble' stage and thereafter	4	Response depends on variety (see p. 110)
Radishes	C	Water to get germination and weekly thereafter	2 (along rows)	Frequent watering will give too much leaf growth
Shallots	C	Water if necessary to get established	Up to $\frac{1}{8}$ pint (70 ml.) per plant per day	Rarely requires watering after establishment
Spinach	A	As frequently as is practical throughout growth	Up to 4 per week	Improves quality

Swedes and turnips	C	Before soil becomes too dry and growth is checked	2 (along row in early stages)	Improves size and quality but reduces flavour
Sweet corn	B	At tasselling and cob-swelling stages	4 on each occasion	Improves size and quality
Tomatoes	B	To get plants established and at flowering and fruiting stages	2 twice per week or more	Very frequent watering increases yield but reduces flavour

* A Crops which respond beneficially to frequent waterings; B Crops which give maximum responses of edible part when watered at the moisture sensitive stages; C Crops which are not so responsive as the others and which should be watered prudently.

5 Insect pests

Nobody likes their efforts to be wasted and that is exactly what can happen in the vegetable garden if steps are not taken against pests. Unless checked, insects can multiply very rapidly when conditions are in their favour. A female turnip moth, the parent of the cutworm caterpillar, can lay more than 1000 eggs. Only one or two caterpillars per square yard can seriously damage potatoes so theoretically only four female moths would seriously affect an acre of the crop. Of course, nature does not allow this to happen. For one reason or another most of the offspring fail to survive. One male and one female offspring surviving will maintain the population of the moth at the same level from one generation to the next. If four survived then the population would double and crop damage increase proportionally.

If anything occurs to favour insects, such as increasing their food supply or eliminating their natural enemies—parasites and predators—then their numbers will increase alarmingly and crops will be severely damaged. However, if conditions become less favourable to the pest then it will decline in numbers, and this is what we must try to do to prevent serious pest damage to vegetables.

Not all vegetables are attacked and some types of pest damage are more serious than others. We need to be able to decide which crops are most at risk, which pests are worth controlling, and how best to deal with these.

Although commercial methods for protecting vegetables on a large scale are often not suitable for the garden, they have evolved after much research and the background information that this has provided is equally relevant to the garden. We now know a great deal about what to do and, equally important, what not to do to protect vegetables against insects and similar pests.

Their life-histories and specific remedies are described in many books and leaflets available to the gardener, but few outline the underlying principles that enable the available methods to be used effectively and their limitations to be recognized.

THE MAIN PESTS

Birds and various small mammals such as rats, mice, voles, rabbits, and hares can damage vegetables as well as the insects and eelworms which will mainly concern us here. The main pests are listed in Table 5.1, which shows when their damage occurs and the more obvious symptoms, as well as indicating broadly the methods available to prevent damage.

Some pests feed on seeds or roots out of sight in the soil, and others on the stems, leaves, or fruits of plants where their damage is more easily seen. Some are single insect species which have to be considered individually, for example the cabbage root fly or the carrot fly. Others can be referred to as groups of species which can be dealt with in similar, though not necessarily identical ways. The aphids are an example. Known variously as blackfly, greenfly, plant lice, dolphins, blight, or merely as 'fly', they are very familiar pests to gardeners. There are many different species of aphids but all have one feature in common: they feed by piercing the plant tissues and sucking the sap. This reduces the vigour of the plant. But they also inject viruses and carry them from plant to plant (see also Chapter 6, p. 168), which is often more important on crops such as lettuce than the feeding damage. However, this method of feeding enables them to be dealt with quite simply by using so-called systemic insecticides that are absorbed by the plants and move with the sap, mainly upwards. The aphids are poisoned as they feed.

Several soil-inhabiting pests feed on a wide range of crops (Table 5.1). The severity of their damage often varies locally from year to year. Examples are the cutworms (caterpillars of noctuid moths), leatherjackets (larvae of the daddy-long-legs flies), wireworms (larvae of click beetles), cockchafer grubs, millipedes (not centipedes), and slugs. Slugs are not, of course, insects, but they can seriously damage many types of vegetables. Most other pests are, however, fairly specific and damage only a limited range of vegetables, usually in the same botanical family. The cabbage root fly only attacks crucifers, which include brassicas and radish, and the carrot fly the umbelliferous crops, carrots, parsnips, celery, parsley, celeriac, and so on. Most aphids are very specific, each species attacking only a very restricted range of host-plants, although the peach-potato aphid is a notable exception and will live on many different garden plants.

It is well worth learning to recognize the different stages of the life histories of pests, each characteristic type of damage and the time when it occurs. If you know what to look for, the pests that affect the leaves, stems, or fruits of vegetables can usually be detected and dealt with before they cause serious damage. Understanding the reasons why pests are tackled in different ways not only begins to make sense of a complex subject but often suggests combinations of methods that are particularly suited to your circumstances and may save a lot of misdirected effort in the garden.

ARE PESTS MORE TROUBLESOME IN GARDENS THAN IN FIELDS?

The answer is very often yes. The very sheltered environment of the garden favours many pests. Gardens contain many suitable host-plants, both cultivated and uncultivated and they are usually flanked by other gardens in varying states of care and neglect. A garden fence is no barrier to most insects and there is little scope for avoiding pest attack by crop rotation on a garden scale. Host crops are likely to be in the ground most of the year, providing ample opportunity for pests to flourish. Furthermore, the actual number of host-plants is usually small compared to the number of young that a single insect can produce. The combination of these factors makes gardens an ecological paradise for many pests and they respond accordingly.

IS ALL PEST DAMAGE IMPORTANT?

Not all pest damage is equally important, so how can we judge when action is needed or when some damage can be accepted? Consider pest damage as three main types. Pests may kill the plant before harvest, they may damage the part that is not to be eaten, or they may damage the edible part. In each case the acceptable level of damage will be different.

If seed, seedlings, or young plants are damaged severely, some may die leaving an uneven stand of plants. The surviving plants then have more room to grow and quite often will become larger than they otherwise would and total yield per unit area may be little affected. This applies particularly to crops growing fairly thickly in rows, for instance carrots, radishes, onions, or peas. Seedling losses of 10–20 per cent are unimportant and hardly noticed, and more often than not these would be dismissed as poor germination. On the other hand a 10–20 per cent loss of widely spaced brassica transplants such as cabbage would be more obvious but may still be often acceptable.

Cutworm caterpillars can also spectacularly kill plants of several crops in this way, including lettuce and young brassicas. They chew through the stems near soil level, either cutting them off or weakening them so much that they are snapped off by the wind. Quite large plants can be killed and a little of this type of damage can be tolerated, but not too much. Death of 20 per cent or more generally calls for action.

Numerous small soil-inhabiting insects can damage seedlings just before or after they emerge from the soil, attacks often occurring suddenly during an occasional warm day in spring. Flea beetles can wipe out a sowing of radishes, swedes, or turnips. A sudden infestation with willow-carrot aphid can kill or seriously check young carrot seedlings. It is therefore essential to watch young seedlings closely to detect this type of damage at an early stage when action is still possible. On the other hand if a severe carrot fly infestation occurs, the maggots can kill even well-established seedlings and nothing can then be done to save them. An insecticide will have to be applied along the rows next time the seed is being sown.

Surprising amounts of damage can be tolerated when it affects only the part of the plant that is not to be eaten and does not kill the plants, especially close to harvest. Complete protection is often not necessary. This applies to cabbage root fly maggots on the roots of Brussels sprouts, cabbages, and cauliflowers during spring and summer, caterpillars or flea beetles eating holes in the leaves of radish (Fig. 5.1(a)), swedes, or turnips, and weevils notching pea or bean leaves, particularly if the plants are growing well. Damage to the roots will only be noticed if it is so severe that the plants wilt or become stunted, or die. With leaf-feeding pests of root crops, you can watch how the damage develops and treat only if the attack is becoming severe. Aphids are an exception because they carry viruses. The peach-potato aphid does not often develop large colonies and is hardly ever serious itself outdoors, but the viruses it transmits

a

b

c

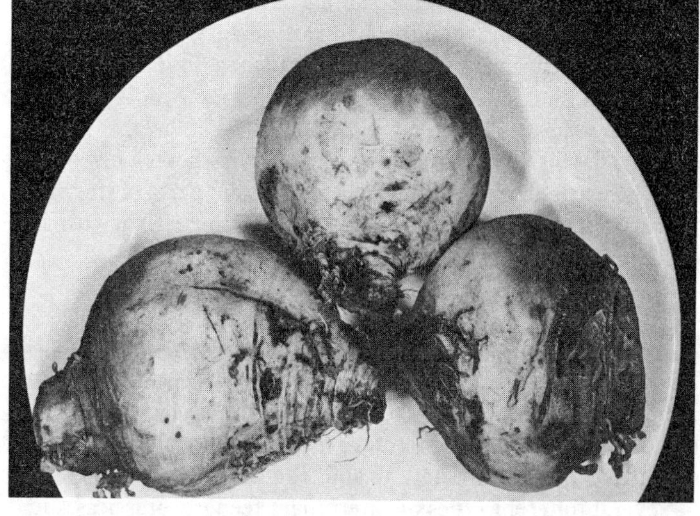

Fig. 5.1. Examples of pest damage that is unsightly but not very important in the garden. (a) Slight flea beetle damage on a leaf of a maturing radish plant; (b) young caterpillars of the small cabbage white butterfly, which can be picked off, and a little frass that can readily be washed off the undamaged cauliflower curd; (c) ugly but largely superficial damage to swedes by maggots of the cabbage root fly.

to many plants reduce their vigour and yields. It is therefore not advisable to allow aphid infestations to persist.

Very little damage is acceptable when pests affect the edible parts of the plants, and then only if it is superficial and there is no intention to store the produce. Apart from the damage, the pest or its debris (frass, cast-skins, or honey-dew) may contaminate the produce often, but not always, making it unsuitable to eat (Fig. 5.1(b)). Often this type of damage does not reduce yield so it is largely an effect on quality of the produce or ease of cleaning it for the table. A little damage by cabbage root fly maggots to swedes (Fig. 5.1(c)) is easily removed when peeling the roots, but in severe attacks the maggots may burrow into the roots and make them virtually unusable. Similarly, carrot fly damage to carrots, parsnips, or celery tends to be very shallow until early autumn, but the maggots then penetrate more deeply and can make crops worthless. Even lightly attacked carrots or parsnips will not store well because the wounds provide entry for bacteria and fungi which cause rots. There will be a lot of wastage unless damaged roots are kept out of the store.

Some damage by pests is tolerable if it only affects a few plants or reduces yield only slightly by affecting the parts of the plant that are not to be eaten. If the edible part is affected, then usually good protection from pests is necessary.

KNOW THE PESTS AND RECOGNIZE THEIR DAMAGE

There are stages in the life histories of most pests when it is difficult or impossible to reach them. At other stages they may be quite vulnerable and therefore it can be important to recognize these and know where to find them.

Several different types of caterpillars feed on brassicas such as cauliflowers, cabbage, and Brussels sprouts, and some have different habits from others. The large cabbage white butterfly lays its eggs in groups of ten to twenty. Its dark caterpillars have gold and green markings and remain gregarious. They can demolish a plant within a few days but often affect only a few plants. The small cabbage white butterfly, however, lays eggs singly on the undersides of the outer leaves of most plants. On hatching from the eggs the pale green caterpillars feed there first and then as they grow older they move towards the centre of the plant. There may only be one or two per plant at any one time. Both of these caterpillars tend to feed mostly during the day. The grey-green caterpillars of the cabbage moth and also cutworms and leatherjackets feed mainly at night and hide during the day. They are easiest to find after dark, using a hand-lamp.

Frequent inspections are important to spot early signs of attack and to note not just whether the plants are being damaged but whether the damage is, or is not, increasing. If it is not getting any worse the attack has probably largely passed, but the weather plays an important role here. Insects are cold-blooded animals and so feed and grow quicker and are generally more active when it is warm rather than when it is cold. Few are active at all when the temperatures fall below about 42°F (6°C) and most reach a peak of activity between 60° and 85°F (15° and 30°C).

In changeable, unpredictable climates there are often seasonal differences of three weeks or more in the times when the pests appear. Calendar dates are therefore not very reliable guides to the appearance of pests and their damage, nor for timing control measures. For example, in the past twenty-five years in central England the first cabbage root fly eggs have been found on dates ranging from 20 April to 8 May. Research studies have shown that the fly begins to lay eggs when the common hedge parsley (*Anthriscus sylvestris*) is just beginning to flower, when about one in twenty of its flower heads are showing white. Brassica plants just being planted or not well established will then almost always need protection against this pest, using either an insecticide or a physical barrier (p. 131) within three days. This is not just coincidence. The cabbage root flies need to feed before they lay their eggs and the nectar of these flowers is often virtually their only source of food in the spring.

Aphid attacks often take gardeners by surprise. These insects can migrate long distances in vast numbers when the weather is suitable, particularly on calm, warm evenings in early summer. They settle out of the air more or less at random and after settling on a plant probe with their mouthparts to determine whether it is suitable as a host before starting to feed and reproduce. Aphids multiply very rapidly indeed at temperatures above about 64°F (18°C), but their colonies can be wiped out even more quickly by heavy rain or fungal diseases. Parasites and predators such as ladybirds, hover fly larvae, and ground beetles prevent rapid growth of aphid colonies, but they cannot be relied upon and often come along too late to prevent serious damage to the crop.

The mealy cabbage aphid first appears on brassicas in early summer, but it does not usually produce very large colonies at this time. Parasites, predators, and fungal diseases appear and the aphid populations usually dwindle and may almost disappear. But do not be lulled into a false sense of security. The worst attacks of this pest usually occur in late summer, August and September. If Brussels sprouts become infested at this time they are difficult to get clean. The aphid secretes honey-dew, which encourages black moulds to grow on the leaves and sprouts. Even if the aphids are killed they cannot be easily removed from within the enfolding leaves of the sprout buttons. Sprouts badly affected in this way are not very acceptable for eating and are unsuitable for freezing.

PEST WARNING SERVICES

At the present time the gardener has to rely largely on his own observations to check the times of appearance of the various pests, but the future should hold brighter prospects. Accurate forecasting and monitoring of pest attacks on commercial crops is becoming more and more essential. Official information services already broadcast warnings and advice on certain pest (and disease) problems on the radio, but visual information on television offers a much more comprehensive possibility for the future. With a special decoding device attached to a television set you can dial a wide range of information displays on many subjects at your convenience. There seems to be no obvious reason why these should not include more pest and disease warnings if farmers and gardeners alike create the demand.

AVOIDING PEST DAMAGE

It is obviously desirable to avoid pest damage whenever possible, but this is only likely to be a complete solution if and when varieties become available that are entirely immune from attack. The only examples of this at the present time are lettuce varieties resistant to lettuce root aphid. Nevertheless there are several other worth-while ways of helping to reduce the intensity of pest attacks in the garden.

Garden hygiene

Good gardening practices have always included keeping the garden clean and tidy. Waste plant material, old plant debris such as cabbage or lettuce stumps, plant thinnings, and unwanted produce harbour pests and should be promptly removed and buried, properly composted, or burned. This helps to deny pests continuity of host-plants, destroys remaining insects and avoids attracting pests from elsewhere.

Much research is now being done to find out what attracts insects to their host-plants and how they recognize them as suitable for egg-laying or feeding. Chemicals emanating from undamaged plants as vapours attract the insects. More vapours are released when plants are cut or bruised. So carrot thinnings left lying on the ground attract carrot fly from greater distances than undamaged growing plants. It will not take the flies long to find the carefully singled plants left in the row and lay their eggs near these.

Some weeds also harbour pests of vegetables, a good reason for keeping down weeds in the vegetable garden, even on uncropped parts. For example, wild radishes can support cabbage root fly maggots and chickweed is a host of the peach-potato aphid and also of viruses that affect lettuce which this aphid can readily transmit. Certain species of thistles (*Sonchus*) can support lettuce root aphid colonies on their roots and provide continuity from year to year.

Resistant varieties

Growing resistant varieties would solve many pest problems. The lettuce varieties 'Avoncrisp' and 'Avondefiance', bred at the National Vegetable Research Station, are so resistant to attack by the lettuce root aphid that for practical purposes they are immune, but this is rare. Research now being done is intended to provide material for plant breeders to develop varieties that are less susceptible to pests than those now available.

There are three main types of resistance to pests. Plants can be tolerant and support large numbers of insects without being severely affected. This is not very satisfactory for vegetables because it permits the insects to do some damage and helps to breed larger populations, making them even more difficult to control. Some plants are less attractive to pests than others, but the most important type of resistance is that where the pest cannot survive as well as on other plants. At its most extreme this type of resistance makes the plants immune to attack, as with the two lettuce varieties just mentioned.

Looking ten years or more ahead, new varieties of such vegetables as carrots, lettuce, and some brassicas should be less susceptible to certain pests, but few are likely to be immune.

Crop rotations

On a garden scale, merely moving crops a few yards one way or the other has little or no effect on attacks by most pests. Only soil-inhabiting forms capable of little movement are likely to be affected, such as eelworms causing various crop 'sicknesses' and, of course, fungi (Chapter 6). To counter eelworms, a three-year system is worth practising, rotating potatoes and tomatoes (hosts of the potato cyst eelworm), brassicas and radishes (liable to clubroot disease and brassica cyst eelworm), and other crops including carrots, celery, parsnips, onions, lettuce, and legumes each of which has other soil-borne pests or diseases. More elaborate systems are possible, but they are not likely to affect the incidence of freely moving insect pests.

Intercropping

Claims have often been made that interplanting crops such as carrots and onions will protect them from attacks by the carrot fly and the onion fly. Research has failed to demonstrate that this is very effective, although it may sometimes result in slightly less damage than if the crops are grown separately. Although not very practical, undersowing brassicas with clover encourages predatory beetles that feed on cabbage root fly eggs and maggots, so helping to reduce attacks by the pest. It also discourages cabbage aphids, but neither pest is entirely prevented from causing damage.

Plant density

Nowadays there is a trend to produce smaller vegetables by growing plants closer together at higher density (Chapter 1). Does this affect pest damage? If we consider a certain number of insects immigrating to attack more plants than usual, they should cause rather less damage to each plant. However, more insects are likely to survive so the pest population in the next generation will probably be larger. The end result may then be little affected when pests such as carrot fly remain local. However, reducing the numbers of plants of a crop between one season and the next may result temporarily in more severe damage. The pest simply responds to the increase or decrease in its food supply, but on a garden scale these effects are unlikely to be great.

Sowing and planting dates

Adjusting sowing or planting dates can avoid the worst periods of attack by some pests. Peas sown early or very late avoid a lot of damage caused by pea moth maggots in the pods. These crops will not be flowering in late June and early July when the moth is laying its eggs.

Very early sowings or transplantings of brassicas pass the most vulnerable stage before severe attacks of flea beetles occur, and late sowings or plantings will also miss most of the attacks.

Carrots sown early will have an extensive root system by the time the carrot fly maggots are feeding on them in June and July and few will therefore be killed, but an early sowing will not prevent them mining in the roots and often killing the tap root. On the other hand, late-sown carrots miss the attack by the first generation of the maggots but are very liable to be damaged in late autumn and winter by the later broods. This pest also attacks parsnips, celery, and several herbs. Growing all of these alongside successions of sowings of carrots virtually guarantees a substantial build-up of the pest and serious damage.

Cultivations

Sometimes it is desirable to sow into a 'stale' seedbed. For example, the maggots of the bean seed fly feed on organic matter in the soil as well as on seeds and seedlings of peas, beans, or onions. It lays eggs freely in freshly disturbed soil and it has been found that delaying sowing for about ten days after the seedbed has been prepared almost entirely prevented attacks on autumn-sown onions. This may also be applicable to peas and beans, which can also be attacked by bean seed fly maggots.

PREVENTING PEST DAMAGE

Methods for preventing pest damage fall into three categories: physical, biological, or chemical. Some examples are discussed here to illustrate principles. There are many old remedies for preventing pest damage, some with little foundation and even in conflict with known facts. The copious use of disinfectant to prevent cabbage root fly damage to brassica plants is an example where research has been unable to support claims made for the treatment. The studies even suggested that the treatment reduced the growth of cauliflowers. Without doing scientifically rigorous tests, it is easy to become wrongly convinced that certain practices are beneficial.

PHYSICAL METHODS

The removal by hand of foliage-feeding pests can be very effective when not many plants are involved. Brassicas can be kept free from serious caterpillar damage in this way. It is best to examine first the underside of the outer leaves where the youngest caterpillars feed. Then examine the centres for signs of their droppings and, if, you see any, search the heart, leaf by leaf, until the caterpillar is found. They are often well camouflaged so you will have to look carefully. Pick it off and destroy it. Finding them takes less time to do than to describe and you should be able to check twenty to thirty plants in ten minutes, quicker than you could prepare a spray and clean the equipment afterwards. Furthermore, you will not have to delay eating the plants to allow the residues of an insecticide to disappear. Examining the plants twice a week in summer can avoid the need for any further treatment.

Occasional aphids can also be crushed by hand to delay the start of an infestation, but it is obviously impracticable to deal with large numbers in this way. An interesting outcome of research into the flight habits of aphids is the use of aluminium foil laid on soil with the shiny side upwards alongside rows of tomatoes or potatoes. The sky is reflected from the foil, discouraging winged aphids from settling on the plants. Apparently they are fooled into flying upwards to find the plant instead of towards the ground.

Brassicas can be protected from the cabbage root fly by physical barriers in the form of either discs of a flexible material placed on the soil surface (Fig. 5.2(a)), or plastic drinking cups from which the bottoms have been removed so that they can be slipped over the plants and lightly pressed into the soil. Both types of barrier reduce the numbers of cabbage root fly eggs laid around brassica transplants and can eliminate up to 70 per cent of the root damage, sufficient for most purposes and, if carefully fitted, virtually equivalent to an insecticide (Fig.

Fig. 5.2. Methods for preventing damage by cabbage root fly maggots to transplanted cabbage, cauliflowers, or Brussels sprouts. (a) A six-inch disc of carpet foam underlay fitted snugly around the stem immediately after transplanting; (b) the maggot damage to the roots stunts the plants (centre) and is prevented as well by a carefully fitted disc (right) as by an insecticide treatment (left).

5.2(b)). The discs should be about 6 in. (15 cm.) in diameter, pierced with a small central hole with a slit to the edge so that they can easily be slipped around the plant stems and lie neatly on the soil surface. A foam carpet underlay makes suitable discs and will expand as the plant stem grows, so eliminating an unsatisfactory feature of the old, stiff, tarred-felt discs used for almost a century. These barriers also act as a soil mulch, reducing loss of moisture. They also provide shelter for beneficial predatory beetles that feed on cabbage root fly eggs and maggots, and this is one reason for their success.

In the future we may see the introduction of traps specifically for some pests. Research is already under way and it remains to be seen how far it will be successful. Traps are available commercially to catch the tiny pea moth to guide the timing of large-scale spraying operations on commercial pea crops, but these are too expensive for most gardeners to contemplate buying at present.

BIOLOGICAL METHODS

The use of biological agents to control pests seems at first sight to be very attractive. However, like the pests themselves, parasites, predators, and diseases caused by fungi, bacteria, and viruses are affected by temperature and are therefore unreliable unless the climate is very predictable. They are often already present in the garden and the problem is how to adjust conditions to favour the natural enemies rather than the pest. In glasshouses this is possible, but we cannot yet do it outdoors on a garden scale.

Bacterial and virus toxins that kill caterpillars are in use commercially in some countries with some success. It remains to be seen whether they become available eventually for other pests and in small packs for garden use.

Are predators and parasites of no use? Undoubtedly they are and without them pest populations would be much larger and the damage to vegetables much greater, although the gardener may not be easily convinced that this is so. The realization of their importance is the basis for sound pest control practice. We may not be able to do much to increase the natural enemies of pests, but we can certainly take care not to destroy any more of them than is essential, if only by being most careful not to over-use pesticides. Ladybirds have been abundant in Europe and overwintered in large numbers in recent years. As they emerge from hibernation sites they can be collected whenever seen and transferred to fruit trees, blackcurrants, and other garden plants where overwintering stages of insects such as aphids are present. Later, putting them on to beans and brassicas when aphids are migrating, or just starting colonies, can very effectively delay the multiplication of the pest. Unfortunately there is no guarantee of success. Only by examining the plants regularly can we decide, for instance, whether natural enemies are holding aphid infestations in check, whether a local thunderstorm has, or has not, dealt effectively with an aphid infestation, or whether a chemical treatment will, after all, have to be used.

CHEMICAL METHODS

There are many old chemical remedies for protecting vegetables from pests, usually aimed at repelling them or attracting them into traps. Some of the chemicals are decidely toxic to man and animals and we would not dream of approving them if they were being introduced as new chemicals today. Soot and whizzed naphthalene may contain carcinogens and sodium fluoride and mercury compounds are acute poisons in their own right. On the other hand, beer is such an effective attractant for slugs that research workers use it as a bait when studying them.

Several well-known insecticides are obtained from plants, for instance derris, nicotine, and pyrethrum. Indeed many plants contain insecticidal chemicals and this may partly account for the fact that most plants resist most insects. This is the basis of many present research programmes to find new insecticides.

Turnips and swedes, for instance, contain certain isothiocyanates while rhubarb leaves contain quite high concentrations of oxalic acid, which is very toxic to man. Since man and the environment has had centuries to adapt to naturally occurring chemicals, long-term pollution problems may be less than with novel synthetic compounds, but this does not apply to their acute toxicities. Nicotine, for instance, is highly toxic to man, and pyrethrum to fish.

Since 1940, chemists have produced many very powerful insecticides but most are not suitable for small-scale garden use. Those available to the gardener in small packs are generally the safest to use, and are unlikely to leave any serious residues in plants or soil, or to affect other animals in the environment when used as recommended. The gardener cannot, therefore, expect to achieve the levels of protection available to the commercial producer.

Even so, a visit to any local garden store shows how much reliance is now placed on chemicals to control garden pests. We need pesticides to garden effectively, but we also need to know how to choose and use them wisely.

PESTICIDES—CHOOSING AND USING THEM

The words pesticides, chemicals, or poisons immediately conjure up emotive thoughts ranging from suspicion and distrust of the unknown to fear bred from the unfortunate influences of fiction. Let us consider a few points that will help to put the record straight. Virtually all chemicals, whether common salt, detergents, therapeutic drugs, herbicides, or insecticides, are toxic to animals or plants if they are present in excess. Some are very much more toxic than others and these we tend to think of as poisons, but it does not make the less-toxic chemicals *non*-poisonous. This is the first fallacy to be dismissed, and leads to an important rule. The amount of chemical reaching an organism determines whether or not it will be toxic. Organisms can tolerate more of some chemicals than they can of others.

Pesticides are toxic to particular pests and diseases, but that does not make them non-toxic to other plants and animals.

There is a second important rule that concerns the time of exposure. The amount reaching the organism multiplied by the time of the exposure determines the toxic dose. Whereas a large dose acts very quickly, a small dose takes a long time to be effective, and very small doses may act so slowly that they do not seem to be toxic at all. The organism can break down and excrete the chemical quicker than it absorbs it so the chemical is ineffective, but that is not quite the same thing as it being harmless.

Most countries now have regulatory authorities who scrutinize and approve new pesticides to ensure that they will not create undue hazards when used (see also Chapter 6, p. 184). They also ensure that the new products will not persist unnecessarily long in the soil or in plants so that the environment will not be at risk, and that they satisfactorily control the pests or diseases as claimed. An immense amount of research and development goes into modern pesticides to make quite sure of these points.

At present pesticides offer the surest means of protecting garden vegetables from pests. If vegetables are being grown for show purposes, and must be absolutely unblemished, then some chemical protection against pests and disease damage is usually essential, but this is by no means necessary if you are growing them to eat. The quest for absolutely blemish-free produce for the kitchen is not only a waste of time and money but it can lead to excessive use of pesticides. The result is attractive-looking vegetables that carry residues of the chemicals well in excess of those considered desirable in food for human consumption. In garden produce you have no way of knowing what residues your vegetables contain. For this reason, if for no other, it is most important to follow carefully the instructions for using pesticides and, if possible, under-use rather than over-use them.

When are treatments necessary? How is their use kept to a minimum? Whenever possible we must try to make use of all available methods and avoid relying entirely on one. This is called 'integrated control'. For each crop consider your past experiences to identify the most serious problems likely to arise and then think of others in relation to these. For example, if birds or slugs regularly take most of your peas then deal with these pests first and the minor damage done by the pea and bean weevil later. Similarly, if every year cabbage root fly maggots kill many brassica plants in the seedbed, then prevent this before being concerned about minor flea beetle damage. If the damage is to the part of the crop you eat, it will of course be less tolerable than if it affects other parts of the plant.

You will need to think ahead (Fig. 5.3). For many soil pests, preventive measures are the most satisfactory and only your past experience with crops in your garden can guide you as to whether they may be necessary. Pests such as cabbage root fly, carrot fly and bean seed fly can be severe but in some circumstances rarely seem to cause serious problems. You are fortunate if this applies to you.

PESTICIDE PRODUCTS

The array of products now available can be quite bewildering. A Directory of Garden Chemicals published in 1977 listed 68 active ingredients in 103 products for pest and disease control. If you already know the chemical required, you need only locate a supplier, but if it is a well-established remedy it is worth while considering alternatives before you buy. Technology is advancing year by year and improved products may now be available. Some shop assistants can guide you in your choice, but reliable informed advice is not always easy to come by.

Whatever your past experience, you will find that the manufacturer's recommendations on the label of the product contain very essential information (Fig. 5.3). Even if you have used the product before, read the label again. Note the nature of the contents, the directions for use, and any safety precautions most carefully. Further information is sometimes given in associated brochures which may be available if you ask for them.

PESTICIDE NAMES

There is much confusion about the names of garden chemicals because each product has two types of name and both are important. The brand name, usually a registered trade-mark, identifies the manufacturer and often, by reputation, the quality of the product. It provides a continuity name for a series of allied products and is usually a guide to reliability. The active in-

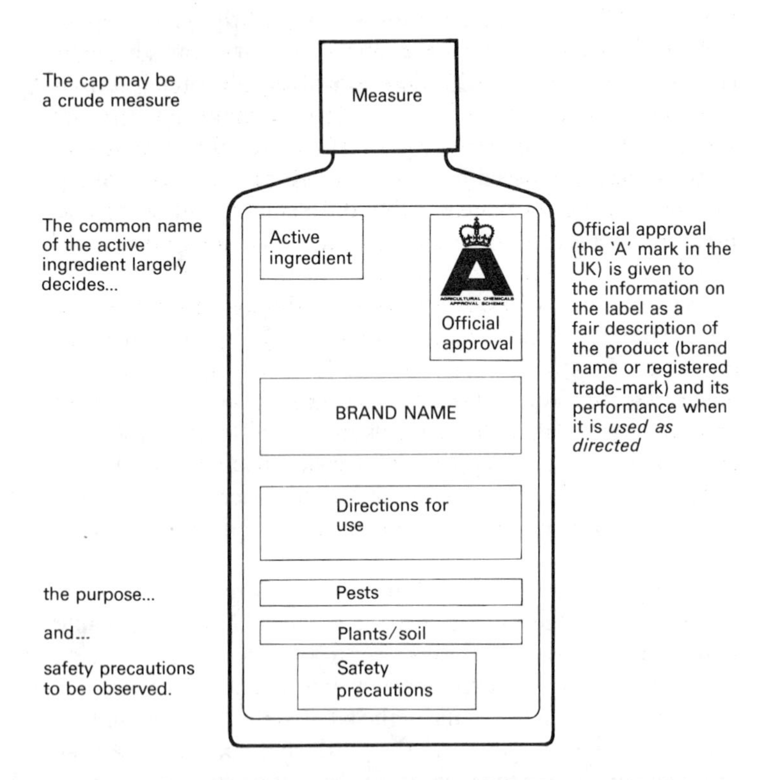

Fig. 5.3. What to look for on the label of a pesticide product. Reputable products and all those officially approved are well labelled and the wording is carefully chosen to mean precisely what it says. Additional brochures giving more information are sometimes also available for some products.

gredient, however, may be changed from time to time even though the brand name is retained, and more than one active ingredient may be present. You will have to search the label well to find out what you are buying.

All reputable products including those bearing a sign of official approval ('A' in the U.K.) will state the internationally agreed common name or names of the active ingredients. Usually this will be in small print, as 'contains dimethoate', but some countries also require a concentration of active ingredients to be stated. This common name is the main clue to the purpose and use of a product. Each active ingredient has particular characteristics that determine how it will behave when it is used and the pests that it can control. It may move systemically through a plant, or it may be particularly suited to use in soil.

More than one manufacturer may sell products containing the same active ingredient, with different brand names and directions for use. The detailed instructions relate to the brand name rather than to the active ingredient.

Some products are available in several different formulations, such as dusts, granules, wettable powders, or liquids. Dusts are mainly used for older materials and few modern products are formulated in this way. They are very difficult to apply precisely without overdosing both the plant and the soil. Dusts are, however, one of the simplest ways of incorporating insecticides into soil and will probably continue to be used for certain products. Granular formulations are intended primarily for application to soil. Again they are not easy to apply precisely, but they are less hazardous to use than dusts. Most garden products are available as liquid formulations which can be used to make sprays or drenches. Sometimes they can be damaging to plants and it is most important to note any cautionary statements on the label indicating the plants which may be susceptible to particular formulations.

Brands containing more than one active ingredient are generally intended to deal simultaneously with several pests. Those claiming to be mixtures for general usage on the blunderbuss principle are convenient, but unfortunately their use conflicts with sound practice in the vegetable garden, since one or other of the ingredients will almost always be applied unnecessarily.

DIRECTIONS FOR USE

The labels on containers are precisely worded to provide reliable information obtained after extensive research and development. It is most unlikely that you can improve a product's effectiveness safely by using your own initiative and ignoring the instructions. Furthermore, you may well create a risk to your crops, to animals, or even to people, including yourself, your family, and pets. As well as telling you what to do, the instructions may also indicate what you must not do, and this is just as important.

The directions for use may indicate when to use the product and will often state how to apply it at the recommended dose. This is usually quite critical and should never be exceeded *in any circumstances.* If you apply excessive amounts, or use too high a concentration, plants may be damaged or the crop become tainted and you may leave excessive residues in the plants or in the soil. Sometimes a slightly smaller dose, perhaps half that recommended, would be sufficiently effective because the manufacturer has to cater for the worst circumstances and these may not be yours. You will have to judge whether the performance of the reduced dose is still adequate, but do not blame the manufacturer if it is not. To reiterate, *never* exceed the recommended dose; nothing is to be gained and harm may result.

A major difficulty arises when very small amounts of diluted sprays or drenches are needed, as in the average garden. Instructions may be given on the amounts of a product to use in 1 or 2 gallons (approximately 5 or 10 litres) but often only 1 or 2 pints ($\frac{1}{2}$ or 1 litre) at a time is needed for a small hand-sprayer. After all, plants do not have to be drowned by spray; in fact most adheres if the spraying is stopped just before the liquid begins to run off the leaves. To mix small amounts it is simplest to abandon fluid ounces or other Imperial measures and deal only with the metric units, millilitres (ml.) and litres. To convert from Imperial units, substitute 28 ml. for 1 fluid ounce, 1·1 litres for 2 pints, and 4·5 litres for each gallon.

Suppose a recommendation is to put 1 fluid ounce of product in 2 gallons of water, and only 2 pints are needed. Two pints is $\frac{1}{8}$ of 2 gallons and so $\frac{1}{8}$ of 1 fl. oz. is required. This is $0{\cdot}125 \times 28 = 3{\cdot}5$ ml. of product for 2 pints of water. A safe and accurate way of measuring out 3·5 ml. is to use a 5 or 10 ml. capacity calibrated plastic hypodermic syringe, marked indelibly with a few spots of cellulose paint to identify it for garden use only. It is also advisable to destroy the needle and in its place attach about 4 in. (10 cm.) of *narrow*-bore flexible plastic tube. This will reach into small narrow-necked bottles used to dispense pesticides and you can easily measure out 3·5 ml., or even smaller quantities. If two or more mixes have to be prepared in quick succession, do not wash out the syringe after each one but be very careful to do so thoroughly when the job is finished. It is essential to wear some form of eye protection and rubber gloves when handling concentrated pesticides, and if there is any risk of getting your everyday clothes contaminated, wear a light overall which can be laundered separately. Whatever you do, do not allow either the concentrated or the diluted material to fall on your skin or to get in your eyes or mouth. If it does so it should be rinsed off immediately.

Products in powder or granular form are very difficult to measure accurately for small-scale use. It is usually sufficiently accurate to weigh out a larger amount (but not on the kitchen scales), and from this to get smaller amounts by carefully heaping it into a cone and halving and quartering until the desired fraction remains. If you can then find a small tube or other container that will hold this quantity, mark it at the right level so that it can be used to measure out further amounts by volume. The amount of granular products to be applied to the soil is often quite critical and the right amount is much less than is generally realized. Sometimes doses of granular and powder formulations are not very clearly stated and you should then determine how little you can use rather than how much, and record the information for future reference.

SAFETY

When pesticides have been used as recommended, they have caused no deaths in the United Kingdom for over twenty-five years. Such deaths as have occurred have been entirely through misuse and the same applies to many other countries. There are a number of important safety precautions to observe to protect yourself, children and others, particularly from the concentrated materials (see also Chapter 6, p. 182).

* Buy no more pesticide than you will use within a season.
* Never keep or use products more than three years old.
* Store pesticides on a high shelf, preferably in a locked cupboard well away from children and pets.
* Wear eye protection and unlined plastic or synthetic rubber gloves when handling the concentrated material, washing them thoroughly inside and outside after use.
* Promptly wash off any pesticide falling on your skin.
* Never suck up concentrates by mouth into tubes or pipettes.
* Do not remain exposed to pesticide vapours for more than a few minutes at a time and then only when unavoidable.
* Follow instructions on the label meticulously.
* Do not spray plants in flower when bees are active.
* Do not eat sprayed crops until the minimum interval as stated on the label has lapsed since the treatment.
* Keep concentrated pesticides out of streams, rivers, and ponds. Most of them are very toxic to fish and other aquatic life.
* *Never, never* put pesticides in containers normally associated with food or drink.

Pesticides are a great boon to the gardener when they are used wisely and judiciously as intended. To avoid killing beneficial insects, it is important not to over-use pesticides. A garden which is sterile of beneficial insects is a very unstable ecological niche. In such situations the gardener may become entirely dependent on chemicals and may well find that their effectiveness fluctuates. This is because they are no longer backed up by the natural parasites and predators which play a hidden, but nevertheless important, part in keeping pest populations within bounds.

World-wide, more than 450 pests of crops and of man have become resistant to many different pesticides. We cannot, therefore, rely on any one insecticide to be effective indefinitely and we know that intensive exposure of insects to chemicals almost guarantees that they will develop resistance sooner rather than later. The more sparingly pesticides are used, the longer they will remain effective.

CROPS AND PESTS

Methods for dealing with some of the more frequent pests of vegetables in gardens are only briefly discussed here and summarized in Table 5.1. A simple form of calendar is illustrated in Fig. 5.4 which a gardener can adapt to suit his own pest control requirements. It will act as an aid to memory so that vital operations are not overlooked.

General soil pests

Cutworms, chafer grubs, leatherjackets, millipedes, wireworms, and slugs can attack most vegetable crops. They occur in large numbers under grass or in neglected gardens and usually become less of a problem after the ground has been cultivated for a few years. If they are troublesome, such insecticides as diazinon, chlorpyrifos, or bromophos worked into the soil before crops are sown or planted will usually be effective. Gamma-HCH (previously called gamma-BHC) can also be used but avoid growing root crops for at least two seasons afterwards because they may be tainted or changed in flavour. Potatoes and carrots are particularly liable to tainting. Against woodlice and millipedes, gamma-HCH can be effective, and against slugs and snails methiocarb pellets are generally better than the older products which incorporate methaldehyde.

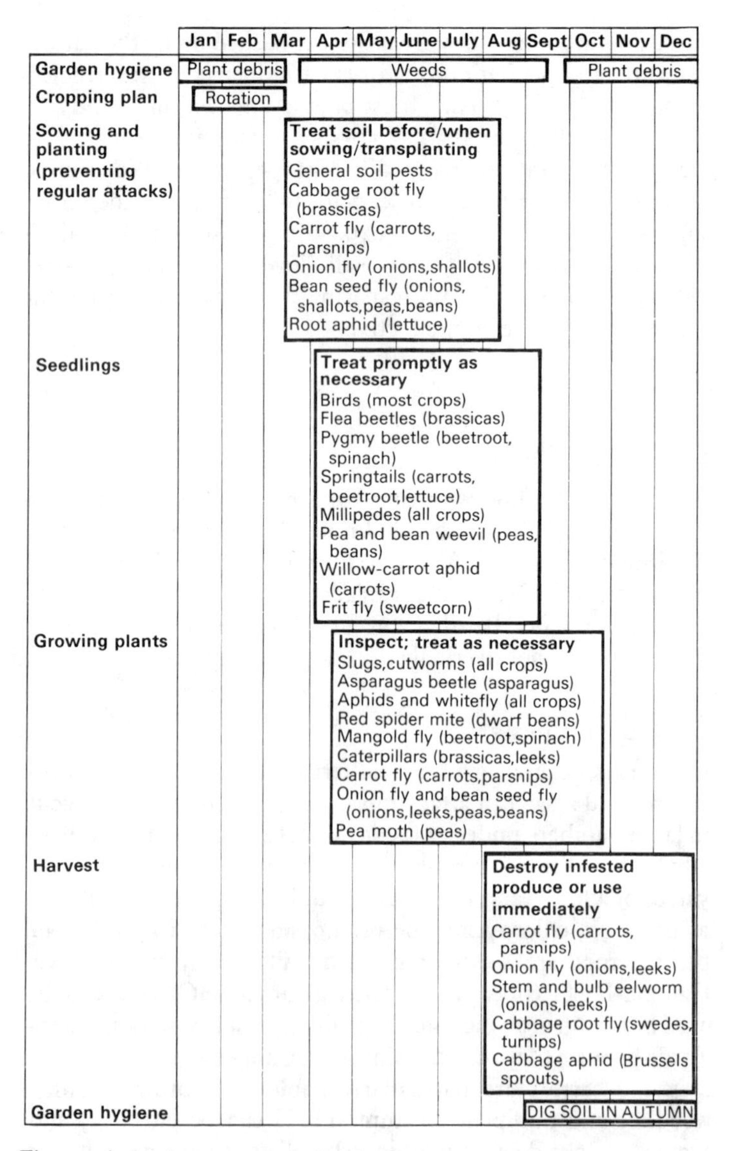

Fig. 5.4. A simple pest control calendar that you can elaborate to suit your own requirements.

Asparagus

The principle pests are *asparagus beetle* and *slugs*. The beetles are brightly coloured, with a reddish thorax and yellow rectangular markings on black wing cases. They appear in early summer and feed on the foliage and stems of the asparagus plant. Their damage can be prevented by treating with derris or malathion when the pest appears.

Slugs characteristically leave shiny slime-trails and can be controlled by slug pellets placed near to the plants.

Beans

The *bean seed fly* lays eggs in freshly disturbed soil and the maggots will feed and damage the seeds and seedlings of all types of beans. If the seeds germinate the plants may be distorted. Where the pest is prevalent, treat the soil along the rows with bromophos, chlorpyrifos, diazinon, or gamma-HCH using either dusts or granules, but ensure that the seeds are sown into a good seedbed so that seedlings grow rapidly and quickly pass the susceptible stage.

The *black bean aphid* infests all types of beans. If broad beans are attacked, allow small colonies to develop for a short while on the tops of the plant to ensure that the immigration of the winged forms is largely over. Then pinch out the tips when the plants are in full flower. If colonies continue to develop, spray with an aphicide such as dimethoate, formothion, or malathion, but not while the beans are in flower or they may harm the pollinating insects. If treatments are necessary while the beans are in flower, pirimicarb can be applied just before dark.

Red spider mites sometimes infest French and runner beans but rarely broad beans. Sprays of dimethoate, formothion, or malathion can be effective, but several applications may be needed at seven-day intervals. If the pods are ready for picking, derris can be used instead.

The *pea and bean weevils* notch the leaves of broad beans and peas, but it is most unusual for their damage to justify treatment.

Beetroot and spinach

Numerous small soil-inhabiting insects can reduce the germination of these crops, but this is rarely very important in the garden. The maggots of the *mangold fly* mine inside the leaves leaving yellowish blisters, but very severe attacks of this pest are necessary before there is any significant effect on yield. Badly infected leaves can be picked off and destroyed and only very heavy infestations need be treated by spraying with dimethoate, formothion, or trichlorphon.

Brassicas and radish

Many different pests attack these crops but not all require treatment in all circumstances. The summer brassicas such as cabbage and cauliflower are most affected. Seedlings can be killed and badly checked by the attacks of the *cabbage root fly*. This can be largely prevented by using physical barriers (Fig. 5.2(a); p. 132) or by treating the soil with bromophos, chlorpyrifos, or diazinon granules when sowing. If established plants are affected, the attack can sometimes be limited by drenching the soil with diazinon or trichlorphon. During the summer period transplanted brassicas should be protected within three days of being planted out. Cauliflowers are the most susceptible, cabbage intermediate, and sprouts relatively tolerant of this pest. Calomel dust gives partial control of root fly as well as helping to reduce infection by clubroot disease. Damage to radishes can be partly prevented by treating the soil at sowing time with the insecticides, but sprays have generally been found to be more effective than dusts or granules. On swedes and turnips, late-season attacks by cabbage root fly are difficult to prevent entirely and usually some damage may have to be tolerated.

Cabbage aphid is readily controlled by such systemic insecticides as dimethoate or formothion. Brussels sprouts are particularly liable to become severely attacked late in the season. A careful watch should be kept to ensure that they do not become too severely affected before spraying. These systemic insecticides tend to move upwards through the plant so it is important to treat the lower leaves even though the infestations often appear on the upper leaves.

Cabbage caterpillars: there are numerous species of butterfly and moth caterpillars which attack cabbages, the most important being the large and small cabbage white butterflies, the cabbage moth, and occasionally the diamond back moth. If the plants are inspected regularly during the summer, caterpillars or eggs can be removed by hand, but if severe attacks occur then spraying with trichlophon or derris, or using carbaryl dust, will usually be effective. The new pyrethroid insecticide, permethrin, is especially effective against these caterpillars.

Flea beetles can seriously damage swedes and turnips in the seedling stage, not only by eating small holes in the leaves, but also because their very small larvae feed on the roots. Derris or gamma-HCH dust are usually successful in preventing excessive damage.

The *turnip gall weevil* produces galls on the roots of brassicas which may resemble clubroot, except that they will be found to contain a very small white grub. This damage is usually unimportant.

Carrots, parsnips, and parsley

Numerous small soil-inhabiting insects can affect the germination of carrots, but the main pests are *carrot fly* and *willow-carrot aphid*. For satisfactory control of carrot fly it is necessary to sow at carefully controlled rates (Chapter 1, p. 21) to avoid the need for thinning, to lift early carrots by the end of August and not to sow late carrots until the end of May or in June. Diazinon applied along the rows at sowing time will usually give good protection of seedlings, but it is not sufficiently persistent to protect crops against the severer attacks that usually occur in late autumn and winter. A second treatment with a spray of diazinon in mid-August may help against these late attacks. However, the gardener has no complete answer available against this pest.

Willow-carrot aphid is readily controlled by a spray with a systemic insecticide such as dimethoate, but it is necessary to examine the plants regularly, particularly young seedlings, to note when the infestation occurs. It can happen very suddenly. On parsnips, a different aphid sometimes appears but is rarely very serious.

Celery and celeriac

The maggots of the *celery fly* can leave unsightly mines in the leaves of celery, but they can usually be dealt with by pinching out the affected leaflets. Severe attacks can be controlled by spraying the leaves with dimethoate or trichlorphon.

Slugs can also damage celery, but these can be kept in check by sprinkling methiocarb pellets along the row.

The *carrot fly* is not easy to control on either crop with the chemicals available to the gardener. A drench with diazinon in late summer may sometimes be effective.

Cucumber and marrow

The most important pests affecting these crops are *aphids* and *whitefly*. Several of the pyrethroid type of insecticides can be used, and also pirimiphos-methyl, but gamma-HCH is liable to damage the plants and should not be used on any plants of the cucumber family. Cucumbers and marrows are particularly prone to be damaged by some insecticides and care should be taken when treating them.

Leeks

Leeks are generally free of serious problems but can be affected by *stem and bulb eelworm*, and occasionally by *onion fly*, *thrips* and the *leek moth*. Treatments are not usually needed, but if eelworm is a problem a sound rotation (p. 128) should be practised.

Lettuce

Several species of *aphids* live on the foliage as well as the *root aphid* on the roots. The root aphid can be controlled by treating the soil for about 6 in. (15 cm.) around each plant with diazinon dust or granules at sowing time or shortly afterwards, or by growing resistant varieties (p. 128). To control aphids on the foliage it is preferable to treat early since they are difficult to kill once the lettuce start to heart. Sprays of dimethoate, formothion, pirimicarb, or derris are suitable, but near harvest time only derris should be used.

Lettuce is sometimes affected by *cutworms*, relatively few of the caterpillars causing serious loss of plants. When this occurs, granules of diazinon or bromophos applied when planting will give some protection, but often it is sufficient to make a careful search in the soil around damaged plants to find the caterpillars and remove them before they do more damage.

Onions and shallots

Where *onion fly* attacks occur the soil can be treated at sowing or planting time with chlorpyrifos or diazinon granules. Alternatively, the rows can be dusted with gamma-HCH when the seedlings are at the 'loop' stage, with a second treatment two weeks later. Badly affected onions should be removed and burned.

The *stem and bulb eelworm* lives in the soil and there are no chemicals which the gardener can use to control this pest. Affected plants should be destroyed and other susceptible plants, members of the onion family, beans, peas, rhubarb, and strawberries, should not be grown on the same land for a further two to three years. Weeds such as chickweed, black bindweed, and mayweed are also hosts of this eelworm and should not be allowed to flourish. Lettuce, or brassicas if club root is no problem (Chapter 6, p. 177), can be grown meanwhile.

Thrips occasionally attack onions in hot dry weather and can seriously damage young plants. If control measures are necessary, then malathion is a suitable material to use.

Peas

Pests other than insects are usually the most damaging to peas at the seedling stage, although they can be affected by *bean seed fly* maggots. Birds in particular can be a problem, preventing peas becoming established. The only remedy is either to stretch black thread along the rows or to put a net over the rows.

The *pea and bean weevil* makes U-shaped notches around the edges of the leaves, but these are not serious and can generally be ignored.

Pea thrips also damage a crop and are likely to be serious if they occur in hot dry weather as the pods are forming. Fenitrothion or dimethoate are both effective against this pest.

The most serious pest of peas is the *pea moth*, which lays eggs on the plant. The young, minute caterpillars bore into the pods and feed on the developing peas. Fenitrothion sprays applied seven to ten days after the *first* flowers appear on the peas in the period June to mid-August will usually help to reduce damage by this pest, but the control is unlikely to be complete.

Potato

Slugs, *wireworms*, and *cutworms* are probably the most serious pests in the garden. Slug damage can usually be reduced by using methiocarb pellets in midsummer, but damage by wireworms cannot be prevented once it has started. If this is likely to occur, for example when using land which has recently been under grass, then it is preferable to treat the soil before planting with bromophos or chlorpyrifos or diazinon granules, but not gamma-HCH which will seriously taint potatoes.

In gardens poor yields of potatoes are frequently attributable to the *potato cyst eelworm* which can build up to large populations in the soil. There are at least two species and various races of this eelworm. Some varieties of potato are resistant to the golden eelworm (cysts are at first yellow before turning brownish) including 'Maris Piper', 'Maris Anchor', 'Pentland Javelin', 'Pentland Lustre', and 'Pentland Meteor', but there are no cultivars at present resistant to the white cyst eelworm. Even if resistant cultivars are grown it is essential to prevent damaging populations of the eelworm building up in the soil by strict rotation of crops on at least a three-year cycle.

Sweetcorn

The larvae of the *frit fly* can damage the growing point of the young sweet corn plants and those sown before the end of May are particularly susceptible to attack. Treatment with sprays of fenitrothion, chlorpyrifos, or gamma-HCH dust applied when the shoots emerge can give substantial protection. The plants only need protecting until they have developed five or six leaves after which they become immune.

Tomato

There are few pests of outdoor tomatoes in the U.K. other than the *potato cyst eelworm* which could present problems to the gardener. To avoid this, tomatoes should not be grown frequently on the same land; a rotation should be used similar to that for potatoes.

Table 5.1
Common vegetable pests and their control

Crop	Pest	Times of damage and symptoms	Control measures: Avoidance	Control measures: Chemical (*common name)
All	Cutworms	May–September. Plant stems eaten away at soil level; holes and cavities in root crops and potatoes in late summer	Destroy caterpillars. Search soil near a damaged plant; at night inspect plants and soil with a hand-lamp	Bromophos, chlorpyrifos, diazinon, or gamma-HCH (not for potatoes or root crops) worked into soil (2 in. deep) before sowing or planting
	Leatherjackets	March–July. Stems bitten through at soil surface; ragged feeding on lower leaves	Ensure soil well drained. Clear land and dig before end of September, especially if previously grass	
	Wireworms	March–September. Stems bitten below soil level. Potato tubers with small deep holes	No root crops or potatoes for two years after digging up grass	
	Chafer grubs	July–October. Roots gnawed 2 in. or more below soil surface	Keep garden clean and weed free	

	Millipedes	April–July. Seed, stems of young plants, roots, tubers eaten (not foliage), and millipedes near by	Search soil near damaged seedling and remove the millipedes	Gamma-HCH worked into soil (not for potatoes, or root crops)
	Slugs	January–December. Lower leaves, stems, roots, tubers holed, mainly in damp sites or during and after wet weather. Slime trails present	Trap with beer in shallow dishes. Good drainage	Methiocarb or methaldehyde pellets placed in small heaps
Asparagus	Asparagus beetle	June–October. Foliage stripped and stems damaged. Beetles and greyish-green larvae present	Remove beetles and larvae	Derris or malathion dust or spray

Table 5.1 (*cont.*)

Crop	Pest	Times of damage and symptoms	Control measures: Avoidance	Control measures: Chemical (*common name)
Beans	Bean seed fly	April–October. Seeds do not germinate, or seedlings weak and distorted	Sow into a good, stale seedbed	Apply dust or granules of bromophos, chlorpyrifos, diazinon, or gamma-HCH along rows
	Black bean aphid	June–July. Colonies of blackfly on stems near top of plant	Pinch out tips of plants when flowered and small colonies formed. Transfer ladybirds to plants	Spray with aphicide dimethoate, formothion, or malathion) after flowering complete or with pirimicarb after dark
	Red spider mite	Mid- to late summer. Dwarf or runner beans, rarely on broad beans		Dimethoate, formothion, or malathion sprays
	Pea and bean weevils	April–July. Leaves notched around margins. White grubs may be damaging root nodules		Malathion or derris sprays or dusts (but very rarely necessary)

Beetroot, spinach	Pygmy beetle Springtails, etc.	April–July. Poor seedling emergence, young seedlings breaking off at soil surface, or falling over		Derris or gamma-HCH along rows
	Mangold fly	Yellowish blisters and mines on leaves	Late sowing (mid-May)	Spray with dimethoate, formothion, or trichlorphon only if damage becoming very severe on young plants
Brassicas	Cabbage root fly	May–October. Seedling and transplants dying, stunted, or wilting. White maggots on and in roots, few side roots remaining, swedes and turnips scalloped (Fig. 5.1(c))	Discs around transplants (Fig. 5.2)	Dip transplants in calomel. Apply bromophos, chlorpyrifos, or diazinon granules to soil around plants within three days of planting-out, or along rows in seed beds

Table 5.1 (*cont.*)

Crop	Pest	Times of damage and symptoms	Control measures: Avoidance	Control measures: Chemical (*common name)
Brassicas (*cont.*)	Cabbage aphid	May–November. Mealy grey aphid on leaves and stems. Leaves puckering	Destroy overwintered brassica plants by May. Transfer ladybirds to lightly infested plants	Spray with a systemic aphicide (dimethoate, formothion)
	Caterpillars	May–October. Holes in leaves, plants 'skeletonized'. Frass in or collecting at base of leaves (Fig. 5.1(b))	Inspect plants twice weekly and destroy eggs and caterpillars	Spray or dust plants with trichlorphon, fenitrothion, carbaryl, derris, permethrin (if available)
	Flea beetles	March–September. 'Shot-holes' in leaves (Fig. 5.1(a))	Sow early (March) or late (June)	Spray or dust seedlings with derris or gamma-HCH (Not usually important on large plants)
	Turnip gall weevil	August–April. Large nodules on main tap-root, with white grub (or cavity) inside		Not important unless seedlings often heavily attacked

Carrots	Carrot fly	June–July. Seedlings die. Plants stunted with reddish leaves. Base of tap root of young plants damaged. October–March. Carrot with mines, some with thin whitish maggots inside	Sow well away from shelter or tall plants, thinly and early (mid-March) or late (mid-June) to avoid eggs of first brood. Lift early carrots before September and all carrots, parsnips, celery by November	Apply bromophos, chlorpyrifos, or diazinon granules, some 2–3 in. (5–8 cm.) deep, under seed-row before sowing. Spray with diazinon twice (late August and early September)
	Willow-carrot aphid	Late May–July. Small pale green aphids on foliage, with white cast skins readily seen. Seedlings soon wilt, become very stunted and die	Inspect seedlings daily late May–mid-June	Spray with a systemic aphicide (dimethoate or formothion) as soon as winged aphids seen on seedlings
	Springtails	April–June. Seedlings dying. Bitten at soil level. Small jumping insects on soil. Not readily seen	Water	Apply gamma-HCH dust if damage becoming severe
Celery, celeriac	Celery fly	May–September. Large blister mines in leaves	Reject seedlings with blistered leaves. Pinch out and destroy blistered leaves	Spray with dimethoate or trichlorphon if attack is becoming severe

Table 5.1 (*cont.*)

Crop	Pest	Times of damage and symptoms	Control measures: Avoidance	Control measures: Chemical (*common name)
Celery, celeriac	Carrot fly	July–October. Reddish-brown small holes and mines in stems and crown at or below soil surface		If severe attacks occur regularly, apply drench of diazinon in trench when transplanting
Cucumber, marrow, pumpkin	Aphids Whitefly	June–September. Stunting with leaves mottled, puckered, and brittle. Colonies of aphids on undersides of leaves	Destroy virus-infected (mottled) plants promptly to prevent spread of infection. Brush off and destroy colonies while small	Spray with aphicide (dimethoate, formothion, malathion). Do not use gamma-HCH
	Red spider mite	June–September. Minute orange-red mites and eggs on undersides of leaves. Leaves pale, brownish, brittle		Spray with dimethoate, formothion, or malathion

Leeks	Stem and bulb eelworm	April–June. Young plants distorted, stunted, and some dying. Later, plant stems swollen, thickened and readily rot	Crop rotation	
	Leek moth	April–May. Centre and growing point of young plants damaged by caterpillar		Carbaryl, trichlorphon, or derris may be effective if applied before attack becomes serious
Lettuce	Lettuce root aphid	June–August. Plants wilting and dying. White waxy colonies on roots	Use resistant variety ('Avondefiance', 'Avoncrisp'). Do not grow lettuce continuously on same land	Apply diazinon dust or granules around plants in June
	Foliage aphids	June–October. Colonies on undersides of leaves and in centre of plants	Aluminium-foil mulch may deter winged aphids from alighting	Spray with systemic aphicide (dimethoate, formothion) if colonies developing, or pirimicarb or derris (near harvest)

Table 5.1 (*cont.*)

Crop	Pest	Times of damage and symptoms	Control measures: Avoidance	Control measures: Chemical (*common name)
Onions, shallots	Onion fly Bean seed fly	May–September. Seedlings dying, often in groups	Sow into 'stale' seedbed especially for August sowings. Remove and burn infested plants	If attacks regularly occur, apply chlorpyrifos or diazinon granules along rows, or gamma-HCH dust (sparingly) at 'loop' stage, and again two weeks later
	Stem and bulb eelworm	April–June. Seedlings stunted, later plants thick-necked	Rotation of crops. Good weed control. Remove and destroy infested plants. No other susceptible crops for two to three years	
	Thrips	June–August (hot weather). Leaves silvery and shine in sun. Many minute blackish thrips present		Spray with malathion if damage becoming severe

Parsley	Carrot fly	July–December. Reddish-brown mines in crown of plant, with small, whitish maggots	Lift and burn badly affected plants	Not usually very important
	Aphids	June–August. Small pale-green aphids and whitish cast skins present		Apply systemic aphicide (dimethoate, formothion)
Parsnip	Carrot fly	June–July. Mining on surface of swelling root. September–February. Reddish-brown mining penetrating root, often with small, whitish maggots	Lift crop and clamp before November to miss worst damage	Protect seedlings as for carrots. Spray crowns of plants in mid-August and in early September with diazinon
Peas	Bean seed fly	April–June. Poor seedling emergence, seedlings distorted	Sow into a good, 'stale' seedbed	Apply dust or granules of bromophos, chlorpyrifos, diazinon, or gamma-HCH along rows when sowing
	Birds	February–July. Seedlings pulled out or severely damaged. Pods ripped open	Net or black thread strung over rows	

Table 5.1 (*cont.*)

Crop	Pest	Times of damage and symptoms	Control measures: Avoidance	Control measures: Chemical (*common name)
Peas (*cont.*)	Pea and bean weevil	March–July. Margins of leaves with U-shaped notches		Not important
	Pea aphid	May–July. Large green (occasionally pink) aphid, mainly on young leaves or growing point	If only a few, remove by hand	Spray with aphicide (dimethoate, formothion, malathion, pirimicarb,, derris) if colonies forming
	Pea thrips	June–August. Flowers fail to develop. Pods silvery and distorted. Black or yellow thrips numerous on flowers and pods		Spray with dimethoate or fenitrothion only if thrips numerous and damage increasing
	Pea moth	June–August. Small white caterpillars in pods eating developing peas	Sow early for crop to flower before early June	For crops flowering after early June, apply spray of fenitrothion seven to ten days after *first* flowers appear

Potato	Potato cyst eelworm	April–September. Foliage stunted, weak, and grows slowly. Roots poorly developed. Tubers small. Minute white or yellow-brown cysts, numerous on roots in July	Rotation of crops and sow resistant varieties (e.g. 'Maris Piper'). Do not grow potatoes on infested land for at least six years	
	Wireworms Cutworms	July–August. Holes in tubers	Do not grow potatoes for two years after grass. Heavy rain or watering in June–July reduces cutworm damage, and some varieties may be less susceptible than others	If regular damage occurs apply bromophos, chlorpyrifos, or diazinon to soil at planting
Radish	Flea beetle	March–September. Numerous small round holes in leaves (Fig. 5.1(a))	Sow very early (March). Inspect regularly for damage	If damage becoming severe, dust seedlings with gamma-HCH or derris

Table 5.1 (*cont.*)

Crop	Pest	Times of damage and symptoms	Control measures: Avoidance	Control measures: Chemical (*common name)
Radish (*cont.*)	Cabbage root fly	May–September. Whitish maggots scoring and mining the developing bulb	Sow early (March) or late (June) to miss damage by the first brood of maggots	Incorporate diazinon or chlorpyrifos granules along rows sparingly before sowing, or spray diazinon along rows when seedlings have one rough leaf
Sweet corn	Frit fly	May–June. Growing point wilts and may die. Small white maggot in central shoot	Sow late or transplant after plants have 5–6 leaves	Spray with fenitrothion, chlorpyrifos, apply gamma-HCH dust as shoots emerge
Tomato	Potato cyst eelworm	May–August. Plants stunted, grow slowly and yield small fruit	Crop rotation	
Turnip, swede	Flea beetle	March–September. Numerous small round holes in leaves (Fig. 5.1(a)). Small white grubs on roots of young seedlings	Sow early (March) or late (June) to miss severest damage	If damage becoming severe, dust seedlings with gamma-HCH or derris dust

Cabbage root fly	June–September. White maggots scoring and mining on and in the swelling main root	Sow after May to miss severe spring attack	Bromophos or Diazinon applied as a spray to crowns in late July–August may prevent some late-season damage as root swells
Turnip gall weevil	April–May. Seedling with round galls on tap root at or just below soil level. August–September. Round galls on swelling roots, contains small, white grub		Protect seedlings with gamma-HCH dust applied along rows. Late-season attack usually unimportant on large plants

* See p. 139 and Fig. 5.3

6 Diseases of vegetables

In some years as much as half or more of the vegetables planted in gardens either fail to grow or, if they do grow, produce a yield well below the true potential. Some of the reasons for these losses have been discussed elsewhere, but undoubtedly one of the major contributors is the group known collectively as the '*plant diseases*'.

WHAT ARE PLANT DISEASES?

Plant diseases are caused by *fungi*, *bacteria*, and *viruses*. They all live on or inside the plant, drawing nourishment from tissues whether they be roots, stems, leaves, or fruits. Like the common cold, these diseases are infectious and in suitable conditions they spread rapidly.

Fungi

The fungi are characterized by the production of fine cotton-like threads, one cell thick and usually divided into segments by partitions at regular intervals. Sometimes many threads intertwine to produce solid structures such as the familiar mushroom. Each segment contains essential living material but never the green pigment chlorophyll. Because they lack this they cannot use carbon dioxide in the air as green plants do and therefore they rely entirely upon the host plant for their supply of food.

Many fungi such as grey mould (*Botrytis*) produce chemicals which dissolve the host tissues while others, for example, the powdery mildews, have special sucking organs which they insert into the host and withdraw nutrients.

Most fungi reproduce by producing spores, not unlike minute seeds, which are spread by wind, rain, and other agencies. Others, for example white rot in onions and clubroot in brassicas, produce resting bodies which can lie dormant in the soil until stimulated to grow by the presence of a susceptible host-plant.

Bacteria

These are often regarded as the most primitive form of plant life. They are a varied group of organisms consisting of minute cells which increase in number by dividing into two. Each half then forms a separate plant, grows, divides again, and continues to do this as long as conditions are favourable. Like the fungi, they contain no chlorophyll and rely mainly upon other organisms to provide them with food.

Some bacteria are beneficial and help in breaking down dead matter, while others absorb or fix nitrogen present in the air, so increasing the nitrogen content of the soil. Another group fix nitrogen within the root nodules of members of the pea family.

Bacteria move in the soil water and can also be scattered by rain and wind. The presence of moisture is essential for bacteria to infect plants; this they do by entering the plant through wounds or through pores on the surface of the leaves. Once established, the symptoms are often distinctive and are easily recognized by the production of mucilage which is associated with a softening of the plant tissues and a strong odour.

Most bacterial diseases of vegetables occur as secondary infections following damage caused by other primary agents. Some, for example halo-blight of beans, are primary parasites and are present within the seed where they infect the plant directly and grow with the emerging leaves (Fig. 6.1).

Viruses

Viruses are composed of simple particles, often in the shape of rods or spheres, which are so small that they can only be seen with a special instrument, the electron microscope. This micro-

Fig. 6.1. Halo-blight of beans. Leaf infection.

scope is capable of magnifying the particles by 200 000 times or more. Mosaics and mottles of dark and light green or yellow areas of the leaves are common symptoms of virus infections such as lettuce mosaic and cauliflower mosaic (Fig. 6.2). Most infections cause some stunting of growth, but others can cause excessive growth either of the whole plant or of particular tissues.

Viruses are responsible for some of the most destructive plant diseases and once a plant is infected, it is almost impossible to cure it. They are spread by a variety of means, some by contact on knives and fingers, by leaf hoppers, mites, white fly, eel-worms, and probably most important of all, by greenfly (aphids, Chapter 5, p. 119). Consequently, apart from the need to prevent the direct damage caused by greenfly, it is also important to eliminate this pest as it helps to limit the spread of many virus diseases. In some cases, the virus may be spread by a fungus, for example, big-vein of lettuce, which is carried by a soil-borne fungus. Other viruses such as bean common mosaic and lettuce mosaic are seed-borne.

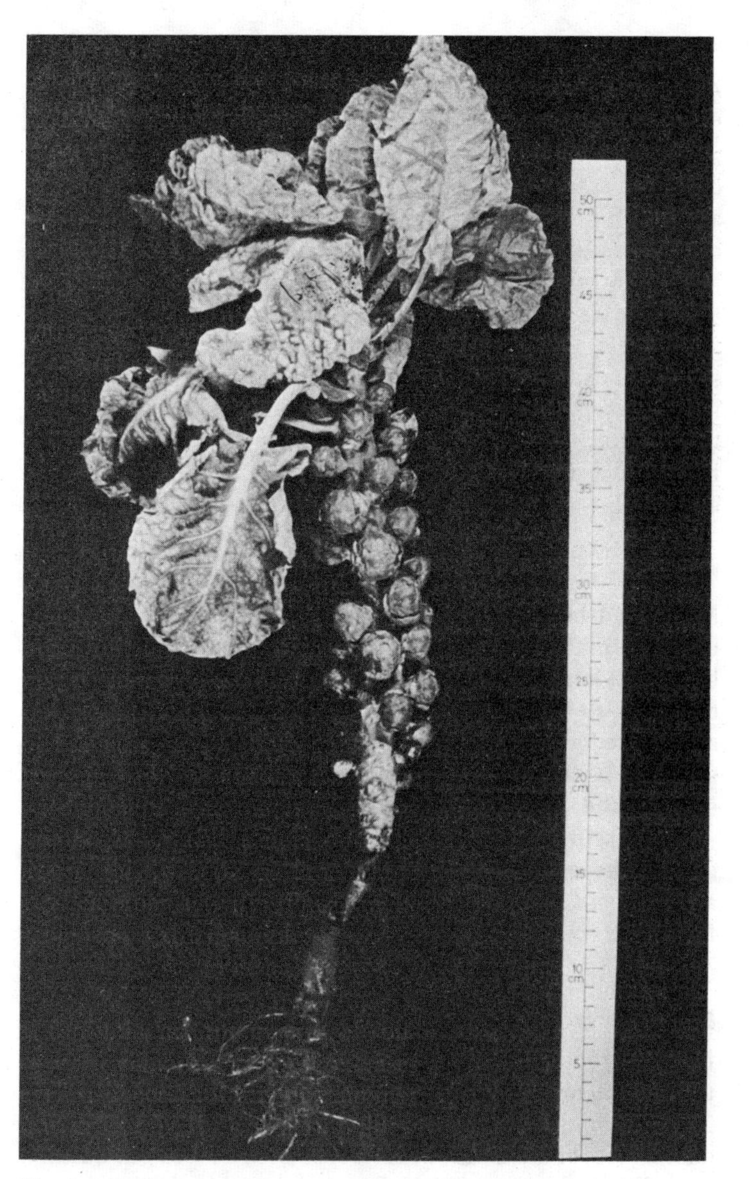

Fig. 6.2. Cauliflower mosaic virus present in Brussels sprouts.

WHERE DO DISEASES COME FROM?

A gardener having decided on the vegetables to be planted for the current season may well ask, 'What is the likelihood of my crops becoming infected by some or all of the main plant diseases.' If his garden plot is situated miles from other vegetables then the chances are very small. However, if the plot is close to a neighbour's vegetables, then the risks become greater.

There are possibly five main sources of infection that all gardeners should consider when attempting to raise disease-free vegetables; these are: wind-blown and splashed-dispersed spores carried in from neighbouring infected plots or vegetable fields, transplants carrying infection either on the plants or in the soil attached to the roots, soil brought in from infected plots, infected debris either introduced or already present on the plot, and finally seeds infected in previous years.

Infection from neighbouring plots

Most wind-blown spores, particularly those produced by the powdery mildews, are released in warm, dry conditions and can be carried long distances by the wind. It is almost impossible, therefore, to exclude this source of infection and the gardener anxious to protect his brassicas or his rose bushes from this group of diseases may have to rely upon the regular application of fungicide sprays.

Those spores which are splash-dispersed and carried in rain droplets have a more restricted range of spread. Diseases such as leaf spot of celery (Fig. 6.3) and black spot of roses tend therefore to be confined to the individual gardens.

Infection from transplants

Plants introduced into new areas will serve as a source of infection if the foliage, stem, or roots are infected. Downy and powdery mildews present on the leaves and clubroot present on the roots of brassicas once established in an area can quickly spread given ideal conditions for growth and multiplication.

Fig. 6.3. Leaf spot of celery.

Infection from diseased soil

As with infected soil present on the roots of transplants, similarly, soil carried into a garden on boots, garden tools, wheelbarrows, etc., can serve to contaminate an otherwise disease-free area with fungi such as clubroot and onion white rot (Fig. 6.4).

Infection from debris

Imperfectly rotted compost often carries grey mould (*Botrytis*) and many of the other diseases previously mentioned. Plants infected in the previous season and thrown on to the compost heap may also carry resting spores of the fungus which are activated in the presence of the new crop. In addition, debris remaining on the ground from the previous year may also be a source of infection.

Infection from seeds

Some diseases such as brassica canker, neck rot of onion, lettuce mosaic virus and halo-blight of beans are able to infect and sur-

Fig. 6.4. White rot of onion. Infected and healthy onions.

vive within the seed. When this seed is planted, the disease organism grows with the developing plant to infect the emerging tissues (Fig. 6.1). It is therefore important to try to obtain seed containing either nil or an extremely low level of infection. For example, with lettuce it is possible to obtain seed containing less than 0·1 or 0·01 per cent mosaic infection. In the event of some of the emerging plants appearing infected, these should be removed as soon as possible to prevent the disease spreading to adjoining healthy plants.

WHAT DETERMINES THE PRESENCE AND SEVERITY OF A DISEASE?

Although a disease may be introduced into a clean area, climatic and soil conditions (in the case of soil-borne fungi) and the growth stage of the plant must be such that they favour the establishment and growth of the disease organism.

Soil conditions

Vegetables grown in a fertile, well-drained soil with plenty of organic matter, a balanced supply of nutrients, and adequate moisture stand an excellent chance of withstanding attacks by most diseases. However, seeds that are slow to germinate because the soil conditions are unsuitable develop an inadequate root system and, as the plants take longer to establish, they may become infected by the damping-off fungi.

In the case of clubroot of brassicas, every effort should be made to avoid growing plants in badly drained areas of the garden as these conditions are thought to favour infection by the motile spores. Furthermore, as an acid soil favours the disease it is important to prevent this by applying lime. A pH of about 7 helps to suppress clubroot activity. This can be checked with pH kits available from garden suppliers.

Climate

Although soil temperatures can influence the growth of soil-borne fungi, the main effect of the climate is on those diseases developing above ground level. In general a warm, dry summer favours the establishment and growth of the powdery mildew fungi whereas a wet summer favours the downy mildews (Fig. 6.5), grey mould fungi, and the leaf-spot diseases. At very low

Fig. 6.5. Lettuce downy mildew. Infected leaf.

temperatures, most fungi and bacteria make poor growth and greenfly do not multiply or fly.

Irrigation

Overhead systems of irrigation often influence the spread of some of the downy mildews and grey mould fungi. If water is applied late in the day the leaves may remain wet throughout the night thereby enabling the spores to germinate and infect. However, overhead irrigation can check the development of powdery mildews.

Planting distance

Densely planted crops can create conditions of high humidity and little wind movement which encourage the growth of many diseases. Close planting can favour disease spread from plant to plant both above and below the ground.

The wider the spacing the less conducive are the conditions for disease establishment and spread. Adequate ventilation at all times restricts most diseases. However, the closer than usual spacings described in Chapter 1 have not unduly increased the risk of disease.

Planting time

In early sowings out of doors, the soil may still be very cold and, as a consequence, seeds germinate slowly and may be quickly infected by fungi and bacteria able to tolerate such conditions. This frequently happens with pea seedlings from early sowings. Had the sowing been delayed until the soil was warmer, the more rapid germination of the seedling would have enabled the plant to escape infection.

If it is necessary to sow early in the spring, then cloches placed over the soil a few days before planting help to raise the soil temperature. In addition, dusting of the seed with benomyl and thiram before sowing will reduce infection.

A firm seedbed with a good tilth favours quick germination, which in turn helps to reduce the chances of infection.

HOW DO DISEASES SURVIVE IN THE GARDEN?

Most diseases are inactive during the resting period of the host-plant. Many fungi survive this period as specialized resting bodies in the soil or on plant debris; others survive as infections in the seed or as fine strands on the surface of the host or in the soil. Most diseases can survive at least one winter, and some, for example onion white rot, have been known to survive for up to twenty winters.

Because viruses spread to all parts of the infected plant, overwintering presents no problems in the case of perennial hosts. Potato viruses can survive in tubers, whereas viruses with a wide host range may pass the interval between crops by infecting susceptible weeds.

Information on these survival patterns is important for it helps to determine when it is safe to replant a particular crop on land known to be previously infected.

ARE ALL FUNGI AND BACTERIA HARMFUL?

Not all fungi and bacteria cause problems as there are some that live entirely on dead and decomposing organic matter. Such organisms are extremely beneficial in that they help to break down waste plant material and so assist in the recycling of nutrients. The rapid removal of plant material, particularly if it is infected with a damaging fungus, can be of great benefit to the control of many diseases. For example, diseases such as grey mould (*Botrytis*) that are able to survive the winter on plant debris would be quickly eradicated if their food source was eliminated. The mildews and moulds on gooseberries, black-currants, and strawberries which survive the winter on fallen leaves are quickly eliminated by fungi, bacteria, and components of the soil fauna which help in the rapid decomposition of the leaf.

Apart from those organisms that live entirely on dead plant material there are others, such as grey mould (*Botrytis*), which can live on decaying matter and on healthy plants; others live mainly on healthy plants but spend a short time in the soil or on plant debris; and finally, there are those fungi, such as the rusts and powdery mildews, which survive the winter in their host-plant without causing appreciable damage.

The latter group, once favourable conditions return, rapidly colonize the host-plant, although usually the degree of infection is insufficient to kill the plant.

CONTROL BY AVOIDANCE METHODS

Some of the precautions that the gardener may take to lessen the chances of infection by diseases have already been discussed in the sections dealing with the source of infections and the factors affecting the severity of attack.

Few gardeners can afford the time or expense of regular applications of chemicals to protect their vegetables. Therefore it is important to consider further, alternative, relatively inexpensive 'avoidance' methods of control which, if properly applied, are very effective.

Time spent in considering points such as the lay-out of the garden (planning crop rotations so that wherever possible 'like crops' do not follow 'like crops'), the careful preparation of the soil, ensuring that the ground is free of infected debris, and checking that planting material is healthy, can all help in excluding many troublesome diseases.

Hygiene

The fungi which cause grey mould overwinter on plant debris; one of the virus diseases, cucumber mosaic virus, that attacks lettuce, vegetable marrow, and cucumber, survives the winter in weed seeds, particularly chickweed. Clubroot survives in the living and dead roots of many brassica weed species. It is important therefore that weeds, plant debris, and roots should be removed and thoroughly composted. In some cases, for example those infected with clubroot, the plants should be burnt. Where material is slow to rot, it should be covered with soil to prevent the release of aerial spores. Plant pots, boxes, and similar containers should be cleaned; water tanks, which may harbour some fungi, should be periodically cleaned. Store rooms, particularly the shelves, should be kept clean and soil and plant debris removed.

Rotation

Rotation of crops is a valuable aid to disease control in that most diseases attack only one crop or group of related plants and become more prevalent when these crops are grown continuously in the same place.

When a particular soil-borne disease such as clubroot (Fig. 6.6) occurs the gardener should not replant that land with brassicas and related crops such as radishes, but with alternative non-susceptible crops such as onions or lettuce.

In some areas repeatedly cropped with a particular vegetable, a condition known as soil-sickness is encountered. In these situations, despite attempts to apply corrective measures, the yield

Fig. 6.6. Clubroot of brassica.

continues to fall. This does not happen with all crops, neither is it always due to the same cause. The decline in pea yields may be due to eelworm presence or to various soil-borne fungi or a combination of both. In other crops, the available nutrients in the soil may have been exhausted.

On a commercial scale the disorder can often be alleviated by fumigating the soil before replanting. Such treatments are believed to be beneficial in that they destroy the injurious organisms that build up over the years and they may also help in releasing nutrients into the soil.

For the amateur there is one sterilant available known as dazomet, but this can only be purchased in commercial packs. Another possible remedy is to replace the diseased soil with fresh healthy soil. While this is a common practice, particularly in glasshouses, in the garden the magnitude of the task, in that it involves the movement of large quantities of soil is such that it often makes this approach impracticable.

Where the disease is known to persist for many years in the soil, for example clubroot of brassicas and white rot of onions, rotation can be of little help to the gardener who has little possibility of moving to an alternative disease-free area. In such circumstances, if these vegetables must be planted, then it is important to apply other control measures (see pp. 189 and 194).

With neck rot of onion, which persists on debris for up to two years, a four-year rotation is an adequate means of avoiding infection of freshly planted seed.

Planning the lay-out of the vegetable garden

When planning the planting schedule, the adjacent crops should always be considered. Where potato blight is a problem, early potatoes should not be planted close to late potatoes, or tomatoes, neither should virus-free seed potatoes be planted alongside old stock infected with mosaic or leaf roll virus.

Seed crops of biennial plants, such as brassicas, should be separated from first-year crops of the same plant or they will probably infect the seedlings with diseases like dark leaf spot, canker, or virus diseases.

Healthy planting material

At all times, it is desirable to start with a 'clean slate'. Having 'cleaned' the land of debris and likely sources of infection, it is sensible to follow by planting material free from viruses, fungi, bacteria, and insect pests. Although it may be difficult to obtain such guarantees, the gardener should attempt to obtain seeds from a reputable supplier who gives some assurance of freedom from disease. But remember, despite these assurances, there is always a danger that some seed may be infected, therefore regular inspection of the emerging crops is advisable followed by prompt removal and destruction of any infected plants.

In the case of transplants, these should be raised in a disease-free area or better still in sterilized soil. If clubroot is suspected in the soil to be used for plant raising, then small amounts may be sterilized by heating for about one hour at 300°F (150°C). This will also kill many other pests and diseases present. Where the number of plants to be grown is limited, raise each plant in an individual pot of sterilized soil and plant out the entire ball. This provides a disease-free base from which the brassica roots can develop. Once the plant is off to a good start, it is more likely to survive subsequent infection by clubroot spores present in the garden soil.

If the garden is free of clubroot then it is a wise policy to obtain brassica plants only from those areas guaranteed free of the disease.

Keeping plants healthy

The use of fungicides applied as seed treatments, dips, or sprays will be discussed in the next section; however, there is much the gardener can do to keep his plants free of disease without necessarily having to resort to the use of chemicals.

The application of a balanced manurial programme (see p. 63), planting at the correct time and depth, and avoidance of overcrowding all assist in creating good growing conditions, so that the plants are strong and more able to withstand infection.

The removal of weeds and of diseased leaves, fruits, or plants as soon as the infection is diagnosed keeps the disease at a very low level and so lessens the chances of an epidemic developing, with the inevitable catastrophic consequences.

The time of the year when the disease first appears is also important. An outbreak of powdery mildew close to harvest time often causes little damage and can, therefore, be ignored, but an outbreak early in the life of the plant can have damaging consequences.

CONTROL BY CHEMICALS

Chemicals should only be used in a small garden when it is considered essential. Chemicals are expensive and, furthermore, with some fungicides the fungus can develop resistance. For example, in certain areas the fungus causing grey mould (*Botrytis*) has developed strains which are tolerant to benomyl, carbendazim, and thiophanate-methyl. In such areas, the disease is not controlled by this group of fungicides and alternative chemicals such as dichlofluanid or thiram have to be used. Therefore, the less the chemical is used, the longer its effective life.

How safe are fungicides?

Although some may be harmful if misused, many are suitable for use in the garden if handled properly and the manufacturer's instructions are *strictly observed* (see also Fig. 5.3, p. 138).

Chemicals are made up in a number of preparations and sold for specific purposes under a variety of trade names. Some compounds are used by seed merchants for treating seeds to eradicate certain seed-borne diseases, for example neck rot of onions, or to give protection against soil-borne damping-off diseases. In this context, it is worth studying seed merchants' catalogues to take advantage of any new developments. Some are used as solutions for dipping roots in before planting out. For example, brassica transplants dipped in a slurry of either thiophanate-methyl or calomel are able to withstand infection from clubroot spores. However, by far the greatest quantity of fungicide is sold for the purpose of spraying plants during the growing season.

With all chemicals there are golden rules which should be observed at all times.

* Handle with care and only use when necessary.
* Keep chemicals and containers away from children and pets.
* Always follow the instructions (never exceed the recommended amount).
* Note the recommended time intervals which must elapse between the last treatment and harvesting; this will avoid harmful deposits remaining on the vegetables.
* Do not spray when bees are active.
* Do not spray when it is windy, as the chemical may drift on to other crops or into your neighbour's garden.
* *Never* transfer chemicals to containers such as beer or soft-drink bottles.
* Wash out sprayers thoroughly after use and further dilute the unwanted fungicide and dispose on spare land in the garden.
* Do not apply fungicides with apparatus which has been used for weed killers.
* Wash hands thoroughly on completion of spraying.

How do fungicides act?

They act either as 'protectants' in that infectious spores landing on the plant surface come into contact with the chemical and are killed, or as 'systemics' which are 'taken up' by the plant. The advantage of the latter group is that there is little danger of them being washed away in rain and, therefore, they remain effective over a long period. Owing to their mode of action, it is often necessary to apply 'protectant' fungicides more frequently than systemic fungicides.

Often a systemic fungicide will help to check a fungus disease that has already become established within the host tissues.

Some fungicides, but not all, can be mixed with other fungicides, insecticides, and foliar feeds; however, *it is always wise to check first with the supplier before using a mixture*, as the combination may be damaging to the plant.

When should fungicides be applied?

Once a disease is established, it is difficult to destroy it although systemic fungicides may check the growth of some diseases. Generally, however, it is best to anticipate the problem and apply the chemical in the early part of the season. Diseases such as blight, leaf spots, powdery mildews, and downy mildews are best prevented by spraying at an early stage of growth before infections have had time to develop.

The choice of chemical is dependent on how it is to be applied, its persistence, will it damage the plant, and whether it will affect the flavour of the vegetables. In the past, a fungicide would control only one disease, but some chemicals, for example benomyl, carbendazim, and thiophanate-methyl, introduced during the past ten years give control of a number of diseases.

It is important, however, to be always on the watch for the first signs of a disease outbreak, *therefore regular inspection of plants is vital.*

The level of control is dependent upon the thoroughness of cover of the plant to be protected. Sprays should only be applied to dry leaves and every effort should be made to wet both the upper and lower leaf surfaces.

Will fungicides damage plants?

The fact that many fungicides persist for relatively short periods does have certain advantages in that there is unlikely to be a build-up of residues in the soil which could be harmful to plants, beneficial insects, and wild life.

Some plants may be damaged by certain fungicides. Seedlings and young plants grown under forcing conditions or in poor light are particularly sensitive. Never spray plants that are wilting because of drought. *In particular, when using a fungicide for the first time on a new variety or unusual plant, check that it is safe by treating a few plants before using the product on a large scale.*

Never hold the spray nozzle close to the plant as the force

of the liquid striking it may be sufficient to cause some mechanical injury which is made worse by the presence of the chemical.

Spraying in hot sun should be avoided as damage due to scorching may occur particularly on young leaves.

What is the best fungicide to use?

Manufacturers of chemicals for both farm and garden use in the United Kingdom can submit their products for approval under the Ministry-sponsored Agricultural Chemicals Approval Scheme. In buying these approved products, the gardener can be certain that the chemicals are of the right kind and in the correct amount for the job they have to do and that the claims for control are fair.

These products contain chemicals which have been considered by the Advisory Committee on Pesticides and Other Toxic Chemicals for possible danger to humans, farm and domestic animals, and wild life, and should not, when used as directed on labels, cause harmful effects. Unfortunately, not all garden chemicals have been submitted for approval; when buying, however, try to obtain approved products distinguished by the mark 'A' on the label (see also Fig. 5.3, p. 138).

Gardeners, other than those in the U.K., should determine if similar schemes are in operation in their countries.

RESISTANT VARIETIES

The stage of development or age of the plant often decides whether or not it will be attacked by a disease. The frequently occurring group of fungi responsible for damping-off, attack in the main, young seedlings and rarely older well-established plants; once the tissues become mature they are usually resistant to infection.

The production of vegetables resistant to all diseases is clearly the ultimate solution to the problems discussed so far. At the present time, there are varieties available that have some if not full resistance to certain diseases. However, there is no guarantee that the resistance is permanent as some diseases are capable of producing new forms that will overcome this resistance. This is seen in lettuces where new cultivars resistant to downy mildew in time become infected with the disease. However, there are many vegetables which have been successfully bred for resistance to specific diseases and included in these is the parsnip 'Avonresister' which is highly resistant to canker.

Even partial resistance is helpful, particularly in localities where the disease is not severe.

BIOLOGICAL CONTROL

The control of plant diseases using other non-damaging fungi, bacteria, and viruses has so far received little attention. It is clear, however, that there are situations where some control is given by these beneficial organisms. For example, many of the powdery mildew fungi are attacked by another fungus and, as a result, the former disease is reduced in severity.

One of the most effective 'biological control mechanisms' is the one whereby earthworms are encouraged to remove plant debris and in so doing remove the host tissue for the overwintering fungus.

Urea applied to fallen apple leaves has been most effective in encouraging a fungal and bacterial population which decomposes the leaf to a stage where it is readily digested by the earthworm. In doing so, the apple-scab fungus which was present on these leaves has been eradicated. Doubtless, similar mechanisms can be introduced into the control of vegetable diseases; the rapid decay of plant material infected with grey mould could be a valuable asset in the control of this fungus.

COMMON DISEASES, SYMPTOMS, AND CONTROL

The following list of some of the more common diseases of vegetables and their symptoms has been prepared as a guide to the gardener in the identification of his problem. In all cases, every effort should be made to prevent the occurrence of the disease by taking 'avoidance' precautions, but where these are impossible or have failed, then the application of suitable fungicides should do much to reduce the severity of the infection.

Although there is a large range of chemicals available for the commercial grower, the amateur gardener's choice is limited, therefore only those that are readily available have been quoted. The fungicides have been identified by the common chemical name of their active ingredient. Manufacturers of fungicides are required to print the common name of the active ingredient on the label, consequently the gardener should have no difficulty in identifying the fungicide.

WHERE CAN THE GARDENER SEEK ADVICE?

Gardeners, like the professional grower, often encounter problems which they cannot solve. In these instances, organizations in the U.K. such as the Royal Horticultural Society, the Agricultural Development and Advisory Service, County Council and Local Authority Advisers, state-financed research stations, and many of the commercial companies supplying chemicals for use in the garden are able to offer advice on identification and the necessary measures for control of the disease. In other countries, such advice may be obtained from similar organizations.

CONCLUSION

Gardeners tend to grow a great variety of crops, each of which may at some time or another suffer from attack by fungi, bacteria, or viruses. Although the number of possible diseases is high, in reality the number likely to be troublesome in any one season is often small. Much of this danger can be further reduced by attention to good management practices and sensible cultivation such as rotation of crops and keeping the ground free from weeds, the reservoir for many of the diseases.

As a general guide to the production of disease-free crops, observation of the following points will go a long way to reaching this desirable 'state of affairs'.

Make sure the soil is well prepared and that the surface is free from old decaying plant material.

Plant high-quality seeds preferably containing full or some resistance to infection. Use certified potato seed; use transplants raised in healthy soil or treated with a fungicide dip; if the disease is one that is known to infect seeds, try to obtain disease-free seed or seed that has been treated.

Plant at the correct time, i.e. when conditions are most favourable for the germination and establishment of the plants. Soils should be warm, moist, but not too wet. A quick and healthy establishment of the seedling will help to protect it from damping-off diseases.

Once the seeds have germinated, thin promptly as overcrowding can lead to stagnant conditions favouring disease outbreaks.

Feed and water correctly.

If disease outbreaks are suspected, apply the relevant fungicides, keeping strictly to the recommendations as laid down by the manufacturers.

Diseased material should be removed and destroyed as soon as possible. If allowed to remain, it will serve to infect the remaining healthy plants.

Non-storable crops, such as lettuce, should be sown little and often. This helps to avoid an excess of overmature plants remaining in the ground. Always remove these plants as they often become diseased and can serve as a source of infection for newly sown plants.

Where possible, rotate the crops as this helps to prevent the build-up of certain soil-borne diseases such as clubroot.

Adherence to these principles, should ensure the production of high quality, disease-free vegetables.

Table 6.1

Common vegetable diseases and their control

Crop	Disease	Symptoms	Control measures	
			Avoidance	**Chemical**
Beans French and runner	Anthracnose (Fig. 6.7)	Cankers on stems, brownish lesions associated with leaf veins. Lesions on pods	Remove infected plants. Plant healthy seed. Rotate crops	Dust seed with benomyl. Spray with benomyl
	Chocolate spot	Brown spots on leaves	Remove infected plants	Spray with benomyl
	Grey mould (*Botrytis*)	Grey mould on pods and stems. Prevalent in wet weather	Remove diseased parts. Ensure good ventilation	Spray with either benomyl, carbendazim, or thiophanate-methyl
	Halo-blight (Figs. 6.1 and 6.13)	Small spots on leaves surrounded by a yellow halo. Water-soaked spots on pods	Remove and burn diseased plants	Spray with copper fungicides
	Mosaic virus	Yellow leaf veins and mottles	Virus is seed-borne. Remove infected young plants early in season and burn	Control of aphids may help to prevent spread of the virus

	Root rot	Leaves yellow, shrivelled. Roots, stem base—brown/black	Remove and burn diseased plants. Rotate crops	Treat seeds with captan
Cabbage, cauli-flower, Brussels sprouts	Cauliflower mosaic virus (Fig. 6.2)	Mosaic with dark green vein banding	Remove and destroy	Control of aphids may help to prevent spread of the virus
	Clubroot (Fig. 6.6)	Discoloured leaves; wilting in warm weather. Swollen roots	Raise plants in sterile soil; keep soil well drained and at pH of about 7	Treat soil with dazomet but allow several weeks before planting. Dip transplant roots in either benomyl, carbendazim, thiophanate-methyl, or calomel
	Damping-off	Plants fail to emerge or seedlings fall over	Raise plants for transplanting in clean soil. Avoid sowing in cold, wet soil	As a precautionary measure dust seeds before sowing with thiram or captan
	Downy mildew	Usually on young plants. Leaves yellow, white areas on under-surface. Often troublesome on cauliflower seedlings	Avoid overcrowding. Raise plants in clean soil	Spray with zineb or dichlofluanid at first sign of disease

Table 6.1 (*cont.*)

			Control measures	
Crop	**Disease**	**Symptoms**	**Avoidance**	**Chemical**
Cabbage, cauli-flower, Brussels sprouts (*cont.*)	Powdery mildew	Powdery white spots on leaves	Remove infected leaves	Spray with benomyl
	Wire stem	Base of stem black and shrunken. Seedlings often die but may survive as stunted plants	Avoid sowing in cold wet soil. Avoid overcrowding	Rake quintozene into the soil surface before sowing
Carrots	Damping-off	Plants fail to emerge	Avoid sowing in cold wet soil	As a precautionary measure, dust seeds before sowing with thiram or captan
	Motley dwarf virus	Centre leaves yellow mottled, outer leaves turn red, yellow, or purple	Remove and burn	Control of aphids may help to prevent spread of virus
	Sclerotinia rot	White mould on stored carrots	Avoid growing on land infected in the previous season	Dust seed with benomyl before sowing

	Violet root-rot (Fig. 6.8)	Slight yellowing of leaves. Roots covered in purplish threads	Destroy diseased roots. Avoid growing root crops on land for at least one year	
Celery	Leaf spot (Fig. 6.3)	Brown rusty spots on leaves and stems	Try to obtain seed treated by the thiram-soak method	Spray with either benomyl or carbendazim at the first sign of the disease
Cucumber family	Cucumber mosaic virus	Mottled and yellowed leaves, surface distorted. Stunted plants may die	Remove and destroy infected plants	Control of aphids may help to prevent spread of the virus
	Damping-off	Plants fail to emerge or soon die	Do not sow in cold wet soil	As a precautionary measure, dust seeds before sowing with thiram or captan
	Grey mould (*Botrytis*)	Grey furry mould on fruit. Attacks stems	Remove and destroy infected parts	Spray with benomyl at first sign of disease
	Powdery mildew	Powdery spots or patches on leaves	Maintain good ventilation	Spray with benomyl or dinocap at first sign of disease

Table 6.1 (*cont.*)

Crop	Disease	Symptoms	Control measures	
			Avoidance	Chemical
Lettuce	Beet western yellows virus (Fig. 6.9)	Interveinal yellowing of outer leaves	Remove and burn. Common in groundsel and Shepherd's purse, therefore keep ground free of these weeds	Control of aphids may help to prevent spread of the virus
	Damping-off	Plants fail to emerge or soon die	Do not sow in cold wet soil	Quintozene dust applied to the soil before sowing helps to prevent infection
	Downy mildew (Fig. 6.5)	Pale green or yellow angular areas on the older leaves; these may bear white spores, especially on lower surface. Infected areas die and become brown. Occurs mainly in autumn	Remove diseased leaves. Avoid overcrowding. Use resistant varieties	Spray with thiram or zineb

	Grey mould (*Botrytis*)	Serious in cool damp weather. Often causes a basal stem rot followed by collapse of the plant. Produces many grey spores on decaying leaves	Destroy affected plants. Remove infected debris. Maintain good ventilation	Spray with either thiram, benomyl, or thiophanate-methyl at first sign of the disease
	Mosaic virus (Fig. 6.11)	Yellow mottling on leaves, growth stunted	Use only seed containing less than 0·1 per cent mosaic infection as stated on packet. Remove young infected plants and burn. Grow resistant varieties if available	Control of aphids may help to prevent spread of the virus
Onions, leeks	Damping-off	Plants fail to emerge	Do not sow in cold wet soil	As a precautionary measure, dust seeds before sowing with thiram or captan
	Downy mildew	Pale oval areas on leaves, tips of leaves become pale and die back. Leaves often fold downwards at infected area	Avoid contaminated and badly drained soils. Ensure adequate ventilation	Spray with zineb at first sign of disease

Table 6.1 (*cont.*)

Crop	Disease	Symptoms	Control measures	
			Avoidance	Chemical
Onions, leeks (*cont.*)	Neck rot (Fig. 6.10)	No symptoms visible in the field, occurs as a grey mould on the neck in store	Remove diseased bulbs from store. Store only undamaged bulbs	Dust seeds/sets with benomyl plus thiram before planting
	White rot (Fig. 6.4)	Plants stunted, foliage turns yellow and wilts. White mould on onion base	Remove and burn. Avoid growing in infected land. Do not delay thinning seedlings	Dust seeds with calomel or apply calomel in furrows at sowing. Spray stem bases with thiophanate-methyl
Parsnip	Canker	Roots blackened and cracked; roots rot	Lime soil, grow resistant variety, such as 'Avonresister'. Later sowings are often less badly affected	
	Damping-off	Plants fail to emerge	Avoid sowing in cold wet soil	As a precautionary measure, dust seeds before sowing with thiram or captan

Peas	Damping-off	Peas fail to grow	Prepare the soil well. Sow when soil warm and not too wet	Dust seeds with benomyl+ thiram to control damping-off and leaf and pod spot
	Downy mildew	Yellow blotches on leaves, brown mould on under-surface	Rotate crops. Remove and burn affected plants	Spray with thiram
	Leaf and pod spot (Fig. 6.12)	Sunken brown cankers on stems of plants, tan-coloured lesions on leaves	Remove and destroy diseased plants	Dust seed with benomyl+ thiram
	Powdery mildew	White powdery spots on leaves. Prevalent in dry seasons	Remove affected leaves	Spray with benomyl
Potatoes	Blight	Appears late in season. Brown patches on leaves. On underside, blighted areas have a white fringe	Avoid planting near discarded potatoes from previous year. Plant healthy seed	Spray with either maneb, zineb, or copper fungicides
	Common scab	Scab areas on tubers. Edges of lesions ragged	Severe on light, dry soils. Dig in compost but do not lime. Use healthy seed	

Table 6.1 (*cont.*)

Crop	Disease	Symptoms	Control measures	
			Avoidance	Chemical
Potatoes (*cont.*)	Leaf-roll virus	Leaflets roll upwards and become hard and brittle. Plants stunted	Remove and burn infected plants. Plant certified seed	Control of aphids may help to prevent spread of the virus
	Mosaic virus	Yellow mottling of leaves	Remove and burn. Plant certified seed	Control of aphids may help to prevent spread of the virus
	Powdery scab	Less frequent than common scab. Lesions are powdery on the surface	Severe on heavy wet soils. Follow a crop rotation	
Tomatoes	Blight (outdoor plants)	Grey/brown edges to leaves; russet-brown 'marbled' areas on fruits. Disease may come from infected potatoes		Spray with either a copper fungicide, maneb or zineb

	Grey mould (*Botrytis*)	Usually occurs on damaged areas. Grey fluffy area	Remove diseased tissues; treat wound with benomyl	Spray with benomyl or dichlofluanid
	Mosaic virus	Light and green leaf areas. Leaf distortion. Stunting or poor growth	Remove infected plants as soon as possible and burn	Control of aphids may help to prevent spread of the virus
	Stem rot	On mature plants. Leaves yellow, brown canker on stem base. Black dots in cankered area	Destroy infected plants. Use sterile soil	Spray stem with benomyl or captan
Turnip, swede	Turnip mosaic virus	Leaf mosaics and/or death of leaf and growing point	Remove and destroy diseased plants	Control of aphids may help to prevent spread of the virus

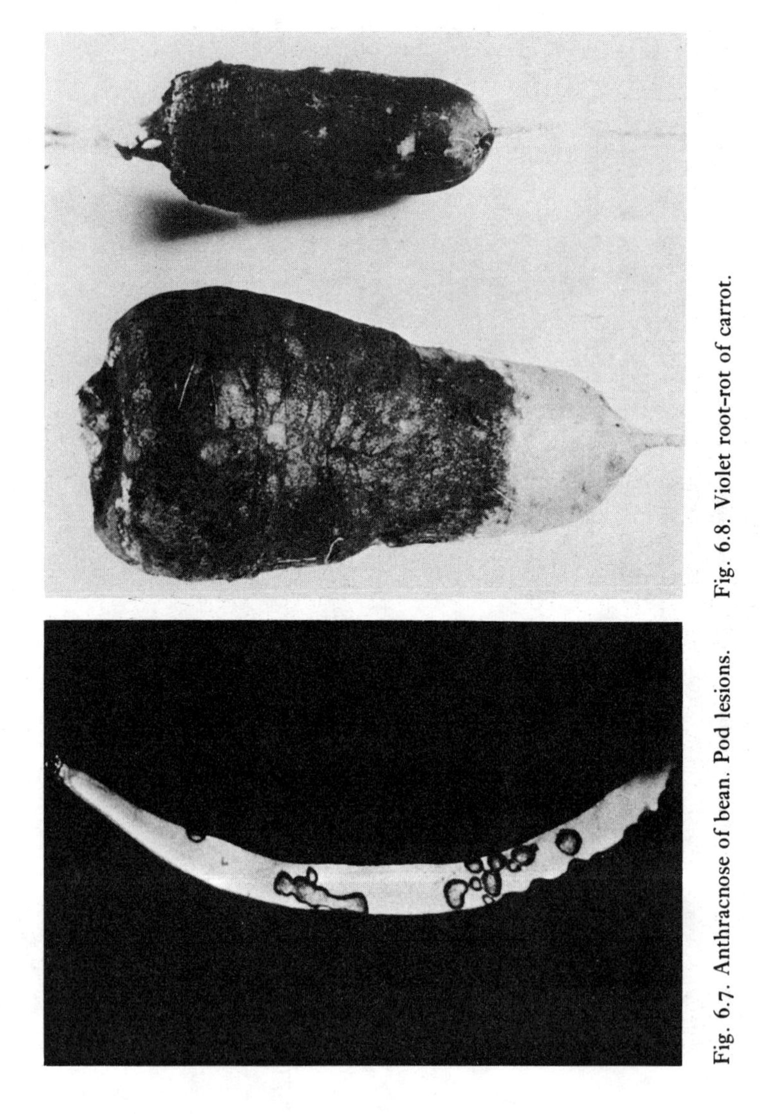

Fig. 6.7. Anthracnose of bean. Pod lesions.

Fig. 6.8. Violet root-rot of carrot.

Fig. 6.9. Beet western yellows virus on lettuce. Infected and healthy leaves.

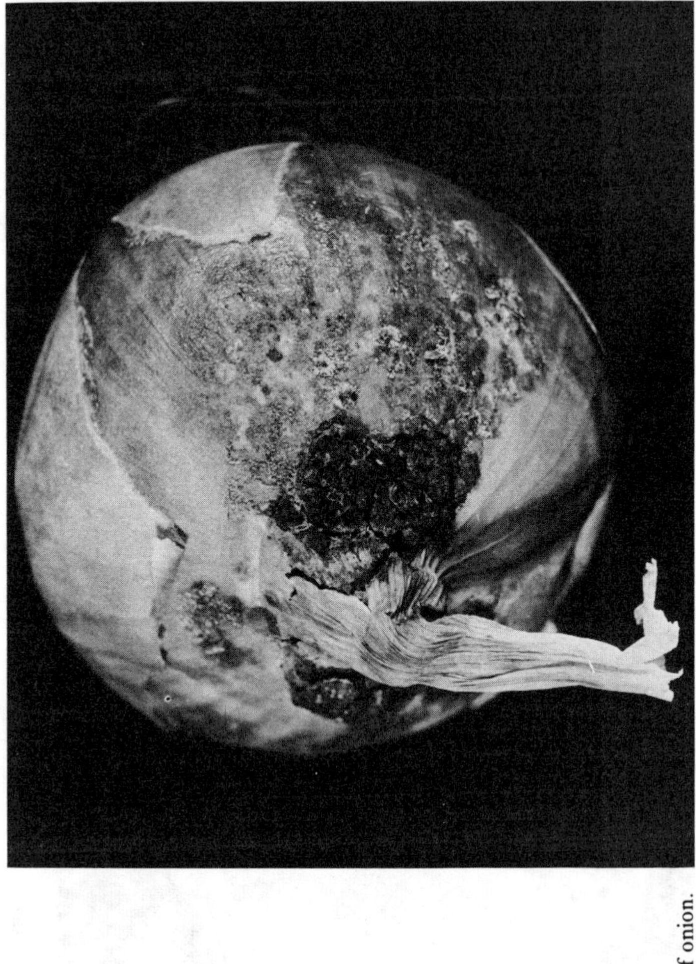

Fig. 6.10. Neck rot of onion.

Fig. 6.11. Lettuce mosaic virus. Healthy (c) and infected (1–5) plants.

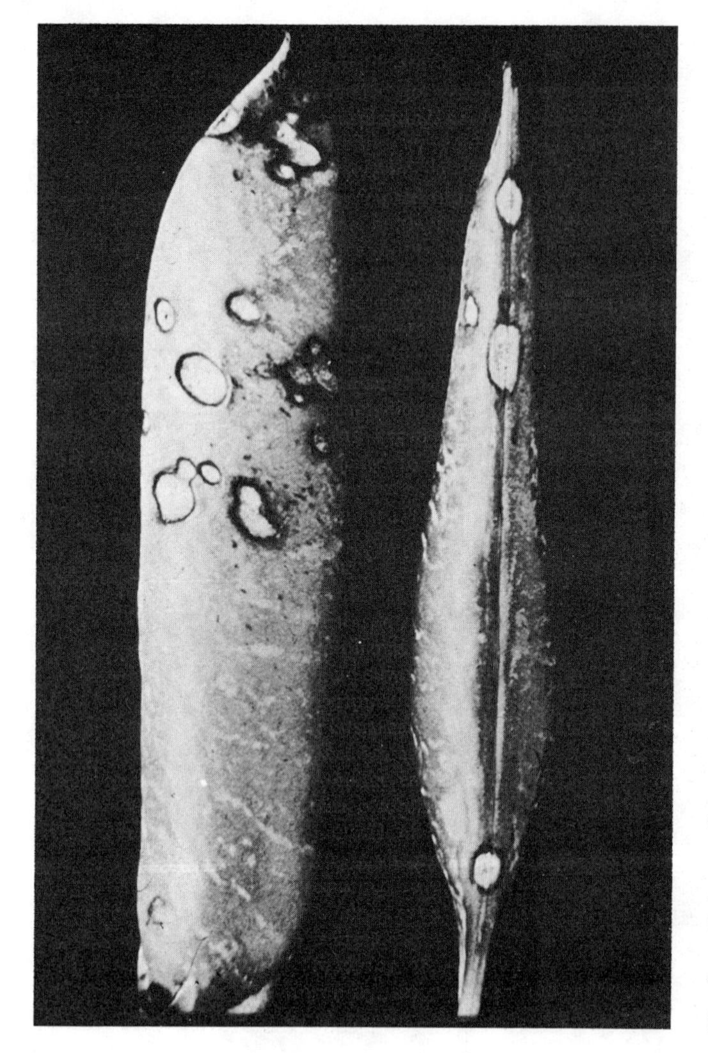

Fig. 6.12. Pod spot of peas.

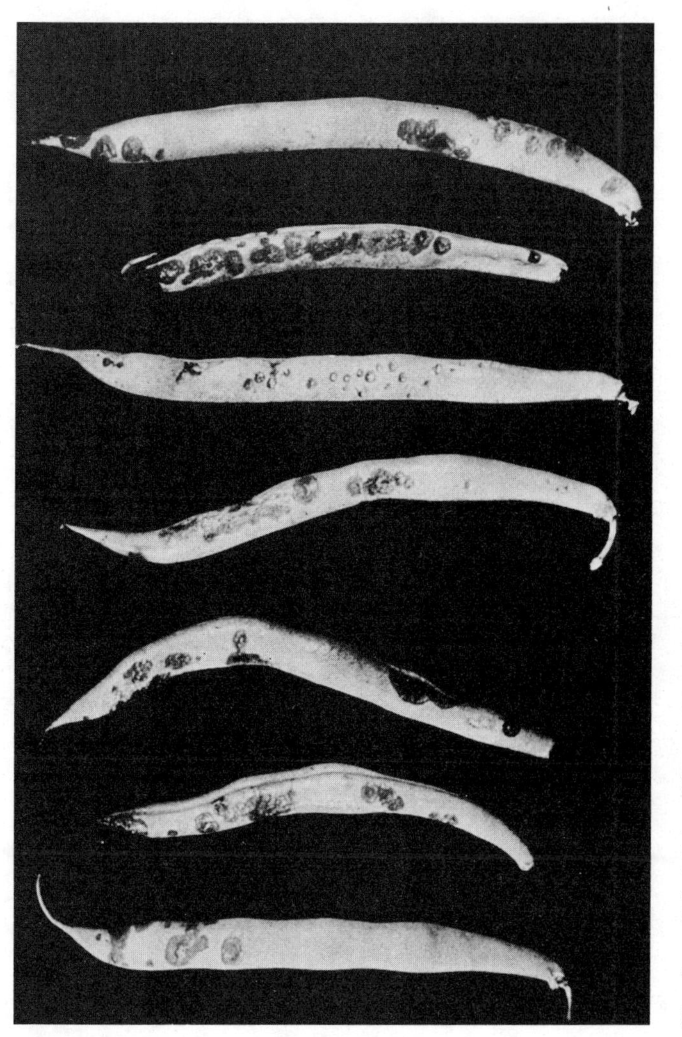

Fig. 6.13. Halo-blight of beans. Pod infection.

INDEX